A potted history of
Fruit

MIKE DARTON

A potted history of
Fruit

a kitchen cornucopia

LYONS PRESS
Guilford, Connecticut
An imprint of Globe Pequot Press

Copyright © The Ivy Press Limited 2011
First Lyons Press edition, 2011

Lyons Press is an imprint of Globe Pequot Press.

Library of Congress Cataloging-in-Publication Data
is available on file.

ISBN 978-0-7627-7060-1

This book was conceived, designed, and produced by
Ivy Press
210 High Street, Lewes, East Sussex, BN7 2NS, UK

Creative Director *Peter Bridgewater*
Publisher *Jason Hook*
Editorial Director *Caroline Earle*
Art Director *Michael Whitehead*
Senior Editor *Stephanie Evans*
Designer *Richard Constable*

Printed in China
10 9 8 7 6 5 4 3 2 1

Color origination by Ivy Press Reprographics

6

Foreword

8

Potted History

A delicious combination of dip-in trivia and essential facts.

*"The trees that are slow to grow
bear the best fruit."*
Molière

118

Table of Latin & Common Names

122

Glossary

126

Index

FOREWORD

HE GROWING OF FRUIT HAS A LONG PEDIGREE. Indeed, in the section on the Fig (see pages 40–41) of this small but densely packed volume we are told that the growing of figs may well pre-date the cultivation of other cereal or vegetable crops by a thousand years. I have been growing fruit, both top (tree-grown) and soft (berries and currants), as semi-standard trees and ornately trained forms, under glass and against walls, for twenty-five years and am well aware of how easy it is to become obsessed with the subject in all its complexity and diversity!

At West Dean Gardens, near Chichester, West Sussex, England, we have an extensive collection of temperate fruits, over 100 varieties of apple, 45 of pear, 25 of plum, 10 of cherry, as well as figs, grapes, peaches, nectarines and the complete range of soft fruit. The point of this collection is to celebrate the long tradition of fruit growing and to educate and enthuse garden visitors about the pleasures of and potential for growing your own, a mission given perfect expression in West Dean's Apple Affair, a weekend pomological fest held in early October. All of this fruit is grown within the sheltering embrace of the flint and brick walls of the 19th-century walled gardens with their range of thirteen glasshouses, including fruit houses. Built into the walls is a circular thatched apple store, an early and elegant precursor to the modern, computer-controlled environment of commercial fruit storage, capable of storing apples and other fruit at the peak of perfection for a number of months.

Part of the fascination of managing such a collection lies in the seasonal round of horticultural endeavor that allows such a collection to flourish and thrive. For these quotidian practicalities my bookshelves contain classic tomes such as *The RHS Fruit Garden Displayed*, *The Lorette System of Pruning*, or the more academic *Temperate-Zone Pomology*. However, equally enjoyable is the enthusiast's desire to visit the more obscure and esoteric shores of the body of knowledge and tradition that complements the practical daily round.

It is this world of fructiferous arcana that *A Potted History of Fruit* is so good at mining and that will yield any number of "Did you know . . . ?" astonishing facts to the fruitarian wishing to deliver the intellectual knockout punch in a duel with a fruit-growing rival. With its attractive format of pithy paragraphs of fruity nuggets alternating with feature spreads on specific fruits, this small volume is ideal bed- or fireside reading that will provide a source of constant delight to any fruit lover. It certainly won't be leaving my bedside table for a fair while yet!

JIM BUCKLAND
Gardens Manager, West Dean Gardens

THE PINEAPPLE
Ananas comosus

BROMELIACEAE

THE PINEAPPLE is the only fruit in widespread cultivation today that is a bromeliad—a member of the family Bromeliaceae, native to the area that is now southern Brazil and Paraguay. Christopher Columbus encountered it on his first trip to the Caribbean in 1492, and thereafter it was brought by the Spanish first to Europe and then to Southeast Asia, notably the Philippines. It was the Spanish who preserved the original Tupi-Guaraní (Brazilian "Indian") name for the fruit, *ananas*, that is now its common name among most of the major languages of the world. At the same time, European Spanish also retains a second name for it that—like English—refers to its pinecone shape: *piña* (hence the drink "piña colada," meaning "strained pineapple").

The fruit itself, technically described as a pseudocarp, actually consists of spiny flowers arranged in two interlocking helixes along an axis—eight in one direction, thirteen at a tangent. Each flower produces its own fleshy fruit tightly pressed against the fruit of flowers next to it. Its fresh juice contains the enzyme bromelain, which is capable of breaking down protein and can, therefore, be used as a meat tenderizer; however, consumption should be avoided by humans who have protein deficiencies or major blood disorders. Otherwise, the pineapple is a good source of vitamins C and B_1 and of manganese. A whole pineapple is sensitive to cold temperatures and should not be stored in a refrigerator.

"The Queen Pine," color lithograph by William Hooker (1779–1832)
Private Collection
WILLIAM HOOKER

TYPES OF FRUIT

There are three major ways in which all the various fruits can be classified:

- ❧ *By scientific genus and species*—in which the plant families Rosaceae and Ericaceae predominate, at least in areas of human habitation.
- ❧ *By reproductive methodology*—as simple fruits, aggregate fruits, multiple fruits, or "accessory" fruits.
- ❧ *By the way they look when dissected*—as berries, drupes, and polydrupes, pomes, and hesperidiums.

To each one of these three lists, however, must be added a category of "others" that do not fit in conveniently. For example, there are still taxonomic problems today in classifying some of the rarer (but edible) cactus and podocarp "fruit." Strawberries may be regarded as "accessory" fruits in that they develop from plant parts other than the ovary, but they are, nevertheless, completely unique in their composition. To accept current definitions of berries means including within that category such "fruits" as tomatoes, avocados, and eggplants, not to mention the watermelon and other melons—and grapes. Meanwhile, a coconut is a kind of drupe, as, indeed, are the olive and the almond; and, technically, the pomes include pears and rowan "berries."

THE ALMOND

Crime writers often tell readers that cyanide smells of almonds, but, in fact, it is the other way around. The almond—like many seeds, especially those of plants in the rose family Rosaceae—smells of its constituent hydrogen cyanide. The almond fruit, a close relative of the peach, is in fact a drupe in which both the outer exocarp and the inner endocarp surrounding the seed have become hard and shell-like. California annually produces eighty percent of the world's almonds.

BANANA JUICE

Unlike most common fruits, a banana has no juice, although the "flesh" of the fruit can be crushed and puréed to a soft, even fluffy, texture. However, the corm— the kind of root structure that continually enlarges and sends up new shoots that together form the "trunk" of the banana "tree"—does contain a juice. It is this juice that is said to be remedial for kidney stones and that, especially with the addition of honey, may be helpful in the treatment of jaundice.

BAN ON GOOSEBERRIES

From the early 20th century until 1966 it was illegal to grow gooseberries anywhere in the United States. That was because the gooseberry bush was considered a potential host for the appallingly destructive fungus known as white-pine blister rust, a danger to the white pine trees that are of enormous commercial value to our country. It was then discovered that cultivated gooseberry bushes constituted little or no risk, and the federal ban was lifted; however, restrictions on their cultivation still remain in force today in some states.

THE STRAWBERRY TREE

"My Love's an Arbutus" (which is not the kind of thing to say to your loved one unless both of you know what an arbutus is, and even then you must be sure that ambient context and temper are right) is the title of an old Irish tune with words by Alfred Graves (1846–1931). It is famous around the world because it is delightfully melodic, yet simple enough to assist in teaching young children how to play the recorder. The arbutus is, of course, alternatively known (in English and in other languages) as the strawberry tree because its fruit look like rounded strawberries, albeit hanging in small clusters on twiglet branches. A statue of a bear eating such fruit appears in the central square of the city of Madrid, Spain, representing a depiction of part of the city's coat of arms. Bears may or may not like it, but the fruit of *Arbutus unedo* is not generally consumed by humans; it is said to be almost tasteless. Nonetheless, in Portugal its juice is distilled into a strongly alcoholic type of brandy that is apparently available under the counter.

LITCHI WINE

The litchi wine that is most popular in China is a thick, golden dessert wine (between thirteen-and-a-half and seventeen percent alcohol by volume) made entirely from litchi fruit. Like most dessert wines, it is very sweet and rich, and for Western tastes, probably best drunk at the end of a meal, when it may also be chilled or served "on the rocks," if deemed appropriate. In China, however, it is frequently used as an accompaniment to meat courses—especially with shellfish or highly spiced meat dishes—and is apparently regarded as a particularly suitable mealtime beverage for women.

THE CHERRY
Prunus spp.

ROSACEAE

THERE ARE fifty-six different named cherry species, and although many of them are native to Japan and the Far East, a surprising number are native to the Himalayan regions. Yet most fresh cherries bought in stores and markets worldwide are examples of various cultivars of either the wild or sweet cherry (*Prunus avium*, also called the mazzard or gean) or of the sour cherry (*P. cerasus*), which are both native to the temperate zones of the Northern Hemisphere, including North America and Europe. The fruit of *P. avium* cultivars can be distinguished from those of *P. cerasus* in that the cherries have a groove partly down one side, whereas *P. cerasus* cherries have little or no groove. Cherries of both species, when fully ripe, come in a range of colors—not just the classic "cherry red"—from deep purple to orange-yellow.

Fruiting cherry trees have a remarkably brief growing duration. The peak time for picking cherries is early to midsummer, which means that cherries are often the first fruits in season.

It should perhaps be noted that the famously spectacular cherry blossom trees that bloom in early spring all over the world, particularly in arboretums in the United States—where there is an annual National Cherry-Blossom Festival—Japan, Canada, and Germany, are *Prunus serrulata* trees (also called *sakura* in Japan), and are not the same as fruiting cherry trees.

"Cherry," color copper engraving by Pierre Joseph Redouté (1759–1840)

from Choix des Plus Belles Fleurs

RETOUCHED BY LANGLOIS

STRAWBERRIES FOR BREAKFAST

During the Middle Ages in northern Europe, strawberries were regarded as an aphrodisiac. A soup made of strawberries, sour cream, and borage was a main element of the traditional wedding breakfast served to newlyweds.

THE DURIAN

The durian is the fruit of trees of the genus *Durio* and is native to Indonesia and Malaysia, although it is also grown commercially in Thailand and the Philippines (both of which countries observe an annual durian festival in specific major cities). On the tree, the fruit looks like an enormous horse chestnut, enclosed in a rounded but spiky husk that may be anywhere from green to brown in color, and can grow up to twelve inches in length and six inches in diameter.

Chiefly because of its size, the durian is known in Southeast Asia as "the king of fruits." Inside the husk, the "flesh," or aril, which is the edible part of the fruit, accounts for only fifteen to thirty percent of the mass of the whole fruit, but it has a smell so pungent that it can be easily detected through the intact husk. That smell is so offensive to so many people— reminiscent, it is said, of a combination of rotting onions, turpentine, and stale cheese—that in many areas of Southeast Asia the fruit is banned from hotels, train stations, airports, and all forms of public transport. Yet the flesh is popular for eating at several different stages of ripeness, and is used to flavor both sweet and savory dishes in Southeast Asian cuisine.

THE OLIVE

The short, gnarled olive tree *Olea europaea* has been subject to cultivation for its fruit—and for the fruit's oil—for at least 5,000 years in the eastern Mediterranean area. The word "oil" (which occurs in variants in just about every language of the world) is derived from the name of the fruit. The olive fruit is a drupe. On the tree the olives ripen from green to purple—but they are mostly harvested in fall or early winter while still green, unless they are to be sold as "natural" black olives, when they are harvested in mid- to late winter. Many black olives in cans and jars are, however, put in their containers when green and are artificially turned black by chemical means (generally by the addition of ferrous sulfate). There are thought to be approximately 800 million olive trees now in cultivation around the world.

HOW TO GROW AN HEIRLOOM CHERRY TREE

Note: The instructions below are for growing cherry trees from pits in traditional home-growing fashion. The eventual fruit—which you must be prepared to wait a considerable time for—is unlikely to taste (or even necessarily to look) like the fruit from which the pits come, because of the horticultural expertise and skills that go into producing table cherries. However, you may be lucky.

1. Soak the pits in lukewarm water and brush each one with a fairly soft toothbrush to remove any remaining "flesh"; rinse them clean and place in a glass or bowl of warm water; let soak for a couple of days, renewing the warm water every twelve hours.

2. Remove the pits from the water and roll them up tightly in a paper towel; put the towel into a sealable plastic bag; seal the bag and put it in the refrigerator for at least three-and-a-half months, periodically checking that the paper towel does not dry out during that time (moisten if necessary).

3. Fill pots at least eight inches deep with equal parts of potting mix and water-retaining granules.

4. Plant your pits to a depth of one-quarter of an inch, and cover with potting mix; using your knuckles, firm the potting mix down; water well.

5. Place the pots on a windowsill that gets a lot of sunshine; allow from three months to one-and-a-half years (depending on the type of cherries you have used—which you may not know) for sprouts to become visible, all the time keeping the potting mix in the pots evenly moist.

6. Plant the sprouts, when they are sturdy enough, to about an inch depth in sandy soil and preferably where they will get a good amount of sunshine; this can be done at any time of the year, but in colder seasons protect them at first with a covering of mesh so that rodents do not dig them up and they are not too battered by wind and rain; water well for the first two years, but thereafter as and when required.

7. As the tree becomes larger, spread organic mulch around the base, and spread an all-purpose high-nitrogen fertilizer in a circle around it (but not touching it); when the fruit start to appear, put up netting to protect them from birds.

Fruit: Do not expect to get edible fruit in less than seven years—and it must be emphasized again that the fruit may taste and look nothing like the original cherries.

THE MULBERRY
Morus spp.

MORACEAE

THE TAXONOMY of mulberries is disputed; well over 100 species of *Morus* have been named, but only between ten and sixteen are generally accepted by all authorities. In practical terms, however, there are only three that are generally regarded as common and of which the fruit are edible, and that are, therefore, commercially grown in temperate and subtropical regions worldwide. They are: the black mulberry (*Morus nigra*), a native of Southwest Asia; the red mulberry (*M. rubra*) of eastern North America; and the white mulberry (*M. alba*) of East Asia. The fruit is not so much a single berry as a multiple fruit, and is unique in shape (although it might be described as a very badly formed blackberry or a highly elongated strawberry). It is baked in tarts and pies, and the juice is made into syrups and wine. Red and black mulberries have a strong flavor; white mulberries are more subtle in taste.

It is the leaves of the white mulberry that are the staple diet of the silkworm (*Bombyx mori*—its own specific name derived from the mulberry tree), whose cocoon has for millennia been used to make silk. Yet it was the black mulberry that was first imported into Britain in the 17th century in a somewhat misguided endeavor to jump-start silk manufacturing there. It soon found an alternative use in folk medicine, notably in the treatment of ringworm (the fungal skin infection otherwise known as dermatophytosis).

"Black Mulberry Tree," by Elizabeth Blackwell (1707–1758)

from A Curious Herbal

JOHN NOURSE, LONDON (1737)

THE MEDICINAL ELDERBERRY

Working on the results of successful experiments to treat the common cold, which were carried out in Israel a decade earlier, between 2002 and 2004 Norwegian chemists proved that extract of elderberry (the berry of *Sambucus nigra*, the elder tree) was not merely effective in protection against both type A and type B influenza but was more effective by quite a margin than all current antiviral medicines normally prescribed for flu (notably for H1N1 flu). This is especially significant in the light of the discovery in 2010 that some flu patients can develop resistance to antiviral medicines within forty-eight hours of their first being administered.

What's more, it is said to be possible to formulate your own preparation of extract of elderberry in the home and at little effort or expense—by steeping dried black elderberries in an airtight jar of vodka and keeping it in a dark cupboard for five weeks, shaking it every few days. At the end of that time (and it is the time that is the main drawback of the suggestion), the fluid can be strained, although that is not essential. The dosage, once the symptoms of flu appear, is one teaspoonful four to five times a day; the symptoms should begin to disappear within twenty-four hours. One teaspoonful per day may act as a preventive measure in flu season.

IS RHUBARB A FRUIT

Rhubarb—which is grown for its stems—is botanically more a vegetable than a fruit. However, a New York court in 1947 formally declared that for the purposes of regulations and duties (and because it was ordinarily used as a fruit) it *was* a fruit.

THE APRICOT DRINK

In markets and stores in the Middle East, sheets or thin slabs of dried apricots are sold wrapped in yellow cellophane, and are especially popular during the month of Ramadan, during which Muslims fast between dawn and sunset and, therefore, concentrate on eating a diet that has to be smaller in quantity, but greater in healthy properties. The apricot sheets are broken up into pieces or strips and soaked in water. The soaking causes most of the apricots to become first a soft mush and then a thick juice. Any remaining hard pieces are strained off, and the juice is sweetened with sugar to taste. This is *amar al din*, otherwise known as "the apricot drink." Dried apricot sheets are also available over much of India, where a similar juice preparation (called *aamsaat*) may be made in the same way, but Indian apricots tend to be far more bitter-tasting.

BREADFRUIT & ANTHROPOLOGY

A large proportion of breadfruit trees scattered throughout the Pacific islands are infertile and seedless hybrids, so the fact that they are so widely distributed in the region can only be explained as the result of deliberate cultivation by human migrants, and particularly by the prehistoric human groups who first colonized the islands. Scientific investigation involving the molecular dating of breadfruit hybrid trees, together with diversely recorded historical and anthropological data, has been able to confirm the overall west–east migration pattern of this early mass human movement from Melanesia toward Polynesia, much of it undertaken by people of the ancient Lapita culture.

GRAPES & WINE PRODUCTION

You might think that the countries that produce the most grapes (viticulture) would also devote the most extensive land areas to viniculture (growing grapes to make wine). That is not entirely the case, as the table below shows. All figures relate to the late 2000s and were published in 2009.

	COUNTRY	VINICULTURE AREA IN SQUARE MILES	COUNTRY	GRAPES PRODUCED (TONS)
1	Spain	4,537	Italy	9,391,050
2	France	3,336	China	7,476,155
3	Italy	3,193	USA	7,037,255
4	Turkey	3,135	France	6,633,362
5	USA	1,602	Spain	6,608,687
6	Iran	1,104	Turkey	3,982,409
7	Romania	958	Iran	3,306,933
8	Portugal	834	Argentina	3,306,933
9	Argentina	803	Chile	2,590,432
10	Australia	634	India	1,838,325

The overall area of the planet's surface devoted to viniculture is estimated to be increasing by two percent annually. Around seventy-one percent of all grapes produced is turned into wine, twenty-seven percent is sold for eating as fresh fruit (of which China produces by far the highest proportion), and the remaining two percent is sold in dried form.

THE WATERMELON
Citrullus lanatus

CUCURBITACEAE

TECHNICALLY, THE WATERMELON (*Citrullus lanatus*) is not a melon but a pepo, a specialized kind of berry, like a cucumber or a squash, so it is not entirely unreasonable to think of it as a vegetable. Indeed, in April 2007 the Oklahoma state senate (somewhat diffidently) passed a bill declaring watermelon to be the official state vegetable. However, although in China watermelon rinds are stir-fried, stewed, or pickled, few Westerners biting into the refreshingly watery, sweet red flesh of a watermelon on a sweltering hot day would consider the watermelon anything other than a fruit. By volume, a watermelon is ninety-two percent water and six percent sugar.

Most fruits in history have traveled from east to west or from west to east. The watermelon is almost unique in having traveled from south to north. It seems that watermelons originated in southern Africa. Yet by the second millennium BCE the watermelon was well established in ancient Egypt; a number of seeds were discovered in Pharaoh Tutankhamen's tomb.

There are currently more than 1,200 varieties of watermelon. Among the latest are the Japanese cubic watermelons, which are grown in glass cubes so that they form in a cubic shape, thereby making them easier to stack and transport. Pyramidal watermelons have also been produced; the challenge is knowing how to cut them into pieces—and they are very expensive.

Citrullus lanatus

"Watermelon," color lithograph by unknown artist (n.d.)
from Le Règne Végétal

RHUBARB

Rhubarb is almost unique among fruits in that it is not actually a fruit at all: what are eaten as fruit in Western countries are the boiled stems (petioles) or stalks of the plant. The plant is actually a relative of buckwheat native to Asia, and of sorrel, which is well known across the Western Hemisphere (and used as a vegetable in West Africa and the Baltic countries Russia and the Ukraine). In Asia, however, and particularly in China, rhubarb's most important form is regarded as the dried rhizome and roots, which because of their unusually high content of anthraquinones can be—and are—used mainly as laxatives.

It was not, after all, until the 18th century that the plant was cultivated for culinary purposes in the West. It is still commonly thought that raw rhubarb stems are toxic to humans—and certainly they are formally so defined in the United States. However, the total amount of oxalic acid in the stems is only between two and two-and-a-half percent of the toxicity of that in the leaves—which are genuinely highly poisonous—and may or may not, therefore, constitute a particularly dangerous hazard. The taste is in any event so strong and acidic that it prevents human consumption.

THE CITRUS FRUITS

Today the most familiar of the citrus fruits—hesperidiums—are the product of human cultivation and hybridization, and the original, natural citrus fruits were considerably fewer in number and generally smaller in size and much more bitter or sour in taste. Modern hesperidiums include:

citron . . . . (*Citrus medica*)
clementine . . . (*C. reticulata* var. *clementine*)
grapefruit . . . . (*C.* × *paradisi*)
kumquat. . (*Fortunella* or *Citrofortunella* sp.)
lemon . . . . (*C.* × *limon*)
lemon, sweet . . . . (*C.* × *limetta*)
lime . . . . (*C. aurantifolia*)
lime, kaffir . . . . (*C.* × *hystrix*)
mandarin orange . . . .(*C. reticulata*)

orange, sour . . . . (*C. aurantium*)
orange, sweet . . . .(*C.* × *sinensis*)
orange cultivars . . . . (bergamot orange, bitter orange, blood orange, etc.)
orangelo . . . . (multihybrid)
pomelo/shaddock . . . . (*C. maxima*)
tangelo . . . . (multihybrid)
tangerine . . . . (*C.* × *tangerina*)
ugli fruit . . . . (multihybrid)

LOGANBERRIES & THE BRITISH NAVY

Because of their exceptionally high content of vitamin C, at the beginning of the 20th century loganberries finally replaced limes (and other fruits) as the primary means for sailors in the British Royal Navy to avoid vitamin-deficiency disorders such as scurvy. For more than ten years, some thirty percent of the vast quantities of loganberries supplied to the navy were produced on a single farm in Worcestershire, England, run by the Norbury family, to whom the famous composer Edward Elgar was piano teacher.

THE SENSITIVE PINEAPPLE

Pineapples are available mostly as canned fruit, in slices or chunks—and there is a reason for this. Fresh pineapples are subject to easy bruising and require extreme care in packing for transportation. Moreover, once pineapples are picked from the tree, they can turn from unripe to overripe within forty-eight hours, and tend to do so independently of others, even from the same batch, at different times. Some consistency of ripening in batches of pineapples can be effected by delicate and potentially expensive temperature control (at the same time avoiding refrigeration), but even this method is unreliable.

PITAYAS

Pitayas, or pitahayas, are fruits of several cactus species native to the Americas from Mexico southward. There are two main genera: *Hylocereus* and *Stenocereus*. The three species of *Hylocereus* are by far the more popular and widespread fruit, known generally as sweet pitayas—whereas *Stenocereus* species are categorized as sour pitayas—and are now cultivated also in a number of Far East Asian countries, especially China and Vietnam, where they are called "dragonfruit."

To eat them, it is necessary to cut through the leathery, scaly outer rind to expose and scoop out the flesh. The red pitaya, *Hylocereus undatus*, has white flesh inside a red rind; the Costa Rica pitaya, *H. costaricensis* or *polyrhizus*, has red flesh inside a red rind; the yellow *H. megalanthus* has white flesh inside a yellow rind. The weakly sweet flesh is eaten raw, together with the small black seeds, which have a slightly nutty taste. Sour pitayas are restricted mostly to arid desert regions of the Americas. Their flesh has a stronger, more sour taste than that of sweet pitayas and is generally more juicy—which is why sour pitayas are a favorite for campers and hikers, besides being an important food source for local populations.

THE ROSE HIP
Rosa spp.

R O S A C E A E

PERHAPS UNFAIRLY, rose hips are known mainly for their herbal properties and not so much for the fact that they taste good. However, maybe that is not too surprising in the light of the fact that rose hips constitute an excellent source of vitamins A, B_3, C, D, and E. In fact, it is because of the notably high vitamin C content that rose hips now tend to be used as dietary treats (if not supplements) for pet guinea pigs and chinchillas. Chinchillas, in particular, have difficulty in metabolizing vitamin C, and rose hips are a convenient way of adding it to their diet without including extra sugar. In a similar fashion, dried and powdered rose hips may be administered at a tablespoonful per day to horses whose coat condition and hoof growth require boosting. In humans, rose-hip tea is said to be especially efficacious in treating bladder infections.

Rose hips do also taste nice, and are used in preserves and jelly, in soups and pies, and in nut bread. Moreover, it is possible to make a wine from rose hips, and in Hungary rose hips are used to make a fruit brandy of the form known as *pálinka*. This is a beverage defined by European Union regulation: it must be made from one hundred percent fruit, with no additives, and be between thirty-seven-and-a-half percent and eighty-six percent alcohol by volume. The strongest *pálinkas* are described in Hungarian as "fence destroyers" (presumably because people who drink the stuff tend to stumble around afterwards and fall over things, breaking them).

"Hairy Dog Rose," color lithograph by J. Sturms (n.d.)
from Flora von Deutschland
K.G. LUTZ, STUTTGART (1904)

KIWIFRUIT PROBLEMS WITH POLLINATION

Perhaps because the cultivated fruit is now somewhat different from the fruit in the wild, the kiwifruit's natural pollinators—bees—seem to take little interest in the flowers, visiting them last and least often, if they bother at all. Many of the commercial growers have to patiently pollinate the flowers by hand using special brushes that puff out air, directing a breeze over the flowers to scatter their pollen. One alternative so far successfully explored is to saturate the entire growing area with beehives so that the bees have little choice but to compete to reach whatever flowers they can get to before the pollen has all gone.

WHAT'S IN A WATERMELON

Just inside the rind of a watermelon, and between the rind and the watery pinkish and reddish "flesh," is a more solid green-white layer that is edible, but that most people tend to leave uneaten. This layer actually contains some nutrients that could be beneficial to consumers. In particular, this layer contains a high proportion of the alpha-amino acid citrulline, which in humans is a substance that contributes to the urea (or ornithine) cycle by which the liver and the kidneys metabolize ammonia to produce urea (which can be excreted in urine).

It is at least partly for this reason that watermelon is mildly diuretic. In fact, if you were to eat more than six pounds of watermelon in one sitting, the effect on your blood plasma could be mistaken for some form of liver or kidney disorder. The Latin name for the fruit is *Citrullus lanatus*, and citrulline is also known to stimulate the production of nitric oxide in the body, which is thought to expand blood vessels—in the same way as does the medicine sildenafil citrate, better known worldwide as Viagra—possibly, therefore, also increasing libido. It is presumably for the same kind of effect that the synthetic salt citrulline malate is now commercially sold as a dietary supplement to enhance athletic performance.

KEEPING APPLES

Apples can be stored for months, and some cultivars for well over a year, especially if wrapped in *Acer* (maple) leaves. However, apples lose their flavor if they are stored unwrapped for any length of time near potatoes, and moreover impart a bitter flavor to the potatoes (and to any carrots that may also be in the vicinity).

IS THE PRICKLY PEAR A REAL PEAR

No, it isn't. Nor is it a fig, despite the fact that another common name for it is the "Indian fig"—which actually corresponds also to its specific name, *Opuntia ficus-indica*. It is nonetheless certainly a fruit, but the fruit of a cactus, and grows at the uppermost tip of a spiny ovoid "arm" or "pad" that projects from the main body of the cactus plant. It has its own smaller spines on its outer surface, which have to be very carefully removed as it is peeled before eating. Although originally native only to Mexico, the prickly pear is now common in the southwest states of the U.S. and in countries with a Mediterranean coastline. Varieties are sold as snacks throughout the eastern Mediterranean region and the Near East.

THE KIWIFRUIT

The kiwifruit (often abbreviated simply to "kiwi"), despite its name, is not originally native to New Zealand but to southern China, which is why its name in English was initially the Chinese gooseberry.

It was only in the early 20th century that Mary Fraser, principal of Wanganui Girls' College in New Zealand, on a visit to mission schools in China, brought home some seeds of the fruit and interested a local nurseryman in growing them. Other horticulturalists in New Zealand began to improve the hybridization and new cultivars appeared. Commercial planting began in the 1940s, which was at the time the only commercial operation in the world to produce the fruit; in China the fruit was traditionally gathered in the wild. Within another decade a name different from "Chinese gooseberry" was being sought, not just to popularize the fruit or particularly for political reasons, but primarily because it didn't look much like a gooseberry and no longer came from China. "Kiwifruit" was finally settled on in the early 1960s. In 2010 New Zealand was the world's second leading producer of kiwifruit, after Italy, but ahead of Chile, France, Greece, Japan, and the United States.

GOOSEBERRY FOOL

"Gooseberry fool" is the English name for a traditional dessert made by puréeing cooked gooseberries in their own juice and folding the purée into a mixture of whipped cream and sugar. Sometimes, rose water is added for extra flavor and a single large (whole) cooked gooseberry is placed on top as decoration. The fool is a form of fruit dessert that dates back to at least the 15th century in England, and fruit other than gooseberries—particularly raspberries and blackberries—have alternatively been common (raw or cooked) in fools since the 19th century. It is apparently called a "fool" because it is light and airy, like the contents of a fool's mind (ultimately from the Latin *follis*, meaning "wind bag").

THE CANTALOUPE
Cucumis melo

CUCURBITACEAE

SURPRISINGLY, THERE ARE two types of cantaloupe melon (or muskmelon or rockmelon), although they are two varieties of the same fruit. The European cantaloupe or Australian rockmelon (*Cucumis melo thelolymphialisis*) has a gray-green skin with somewhat shallow reticulations (ribbed networking), whereas the North American cantaloupe (*C. melo reticulatus*) has a light brown skin with deeper, more marked reticulations. Both have light orange flesh, although the flesh of the North American cantaloupe could in fact be more pink or yellow, except that commercial growers have tried to selectively breed for the orange-fleshed variety.

However, describing cantaloupes as "European," "Australian," or "North American" tends to gloss over the fact that cantaloupes originated in India and Africa. They appeared in Europe only after being imported by the ancient Egyptians. It was Christopher Columbus who introduced cantaloupes to North America on his second voyage to the New World in 1494.

The surface and, once opened, the flesh of a cantaloupe is extremely susceptible to *Salmonella* bacteria. Always wash the outside of a cantaloupe before cutting, and wash the flesh, too, if it has been kept for more than twelve hours, especially if unrefrigerated. Throw away altogether if kept (even refrigerated) for more than seventy-two hours, because if *Salmonella* has not attacked it, other bacteria are likely to have.

"Orange Cantaloupe Melon," color lithograph by Faguet (1841–1886)
from Album Vilmorin
VILMORIN-ANDRIEUX & CIE (1850–1895)

MANZANITAS & BEARBERRIES

Manzanita is the Spanish diminutive of *manzana* or "apple," and is the general name given to the shrubby plants of the genus *Arctostaphylos*. The cranberry-like berries and the flowers of most of the 106 species of the genus are edible. The bark by itself can be used in infusion as a tea. Native Americans formerly used manzanita leaves to clean their teeth.

Bearberries are the three species of manzanita (*Arctostaphylos alpina*, alpine bearberry; *A. rubra*, red bearberry; *A. uva-ursi*, common bearberry) that have over the millennia adapted to Arctic and subarctic conditions (which explains why the English term for them suggests that they come from where bears live). Their leaves may be used in herbal medicine, but it is the berries that have historically been regarded as both edible and curative. A veritable raft of beneficial applications in folk medicine has been ascribed to them (especially by Native Americans, and especially in respect of ailments in the cardiovascular and urogenital departments)—although as yet there is scant scientific evidence to support any such claim. Moreover, it is said that eating large quantities of bearberries may, in fact, cause nausea, vomiting, high temperature, ringing in the ears, bluing of the skin, and discoloration of the urine, and that some people (notably women who are pregnant or nursing) may be physically intolerant of them in other ways besides. Nonetheless, bearberries—particularly the common bearberry, in English also called the "bear's grape" (a translation of the Latin specific name)—have been included in herbal textbooks and rural medicinal lore since at least the 13th century, when they were described as having such uses both in a medical treatise from south Wales and by Marco Polo while traveling in China.

APPLES, ANCIENT & MODERN

The apple has been cultivated by humans for at least three millennia. In fact, apples were cultivated in several different varieties in the Nile Delta by command of the ancient Egyptian pharaoh Rameses II during the 13th century BCE. Some six or seven centuries later, in ancient Greece, apples were so highly valued that when one was tossed by a young man to a willing maiden, it represented the most romantic form of wedding proposal (as long as the girl caught it). The Greek-writing Roman author Pliny the Elder during the first decades CE listed thirty-seven different varieties of the fruit. By 1640, an English horticulturist counted sixty varieties, a total that rose to ninety-two within thirty years. In 1866 the list had reached 643 known cultivars. Today there are at least 7,500, all of which are sufficiently different from the others to be distinguishable by an apple expert.

HOW TO GROW AN HEIRLOOM HONEYDEW MELON

Select a site that gets a lot of hot sunshine, is generally out of the wind, and yet has open space around it; honeydews need at least eight weeks' solid heat and a soil temperature that averages above seventy degrees Fahrenheit—conditions that can, with some artifice, be obtained even in temperate zones. Buy seedlings ready for transplant from a garden center or nursery and plant from mid- to late spring, when any danger of frost is gone.

1. Construct raised, flattened mounds of soil around three feet in diameter and about eight inches high, the bases of which enclose around five or six inches of well-rotted manure or compost; ensure that no mound is within four feet of another; cover each mound with a black plastic sheet securely attached to the ground all around the base of each mound to retain heat.

2. Plant three seedlings per mound, one inch deep, at the points of an isosceles triangle some six inches in from the mound's edge, cutting X-shaped slits in the plastic sheet, positioning the seedlings into the soil, and leaving the flaps in the plastic open; spread the mounds with floating row covers (to retain heat and protect from insects while also allowing moisture through).

3. Keep the soil moist, but not wet. To water, temporarily remove the floating row covers and apply the water directly to the soil (under the seedling or any new growth) at the X-shaped slits; be sure to replace the floating row covers afterward each time.

4. Every three weeks or so, use a high-nitrogen fertilizer until flowers appear; use a high phosphorus-potassium fertilizer after flowers appear, and remove the floating row covers for longer periods of daylight to ensure pollination by bees.

5. When fruit appear, prevent them from resting on bare soil (which might cause them to rot) by spreading a layer of mulch beneath them or lifting them onto upturned plates, flowerpots, or large, flat stones. You may also have to use preventive measures against rodent predation, especially if the leaves are in danger (melon sweetness comes from the leaves). Inspect under leaves daily for aphid and fungal attack (both can be treated).

Harvesting: Cut down on watering as ripeness approaches (to increase internal sweetness); a ripe melon slips straight off the vine and the blossom end should yield a little to mild thumb pressure.

THE ORANGE

Citrus × *sinensis*

RUTACEAE

THERE ARE TWO main species of orange: the sweet orange (*Citrus sinensis*) and the bitter orange (*C. aurantium*). Historically, it would seem that the bitter orange is by far the earlier, probably first cultivated—maybe as a hybrid of the pomelo (*C. maxima*) and the mandarin (*C. reticulata*)—in ancient Southeast Asia. It was the bitter orange that was first introduced to Europe from Persia during the 11th century. Sweet oranges were brought by Portuguese traders from India to Europe as late as the 15th century, and immediately became far more popular. The majority of the oranges cultivated today are, accordingly, varieties of sweet orange.

Two well-known varieties that originated not as hybridized cultivars but by genetic mutation, however, are the blood orange and the navel orange. The flesh and juice of the blood orange have an unusual pigmentation that some regard as reminiscent of blood (although the taste is normally unaffected). The "navel" of the navel orange is in fact a conjoined twin growing at the base of the "host" fruit opposite the stem. Navel oranges are seedless because of this mutation, and are commercially grown using cuttings grafted onto other citrus rootstock. Ripe oranges can be stored at room temperature for up to a week or up to four weeks in the refrigerator. In the refrigerator, however, nearby meats and dairy products may be contaminated by orange-zest vapor.

"Orange," color copper engraving by Pierre Joseph Redouté (1759–1840)

from Choix des Plus Belles Fleurs

RETOUCHED BY LANGLOIS

NOAH & THE GRAPE

After the Flood, Noah was reportedly the first human to plant the vine (Genesis 9:20)—and to enjoy the alcoholic beverage made from its fruit. His subsequent state of inebriation caused irreparable family disruption. However, this personal setback does not seem to have affected his overall health and well-being. He went on to live for at least another 345 years.

THE MELONS OF CHINCHILLA

The town of Chinchilla in Queensland, Australia, produces around a quarter of all the country's melons, including honeydew melons, cantaloupes (also known as rockmelons in Australia) and watermelons. Accordingly, since 1994, the town has been the center for a biannual festival of "all things melon." At the 2009 Chinchilla Festival, which won the Queensland Regional Achievement and Community Award for Tourism, 10,000 visitors were estimated to have attended on the main day alone. Highlights of the festival were such events as melon bungee, melon chariot, melon Ironman, melon bullseye, melon skiing, melon-eating contests, and a melon-seed-spitting competition—evidently it was a very Australian kind of gathering. Another feature of that year's festival was the raising of the world record (as formally attested in the *Guinness Book of Records*) for melon head-smashing—splitting open as many watermelons as possible in one minute with one's head—from forty to forty-seven by one John Allwood.

TUTTI FRUTTI

Tutti frutti is Italian for "all fruits," but on menus in the English language generally applies to a form of (vanilla or vanilla and strawberry) ice cream containing small pieces of candied fruits. Alternatively, it can occasionally mean a small mixed fruit salad used as a sweet course/dessert for a luncheon.

POMEGRANATES & JUDAISM

The pomegranate is regarded in Judaism as a very old symbol of righteousness because, according to traditional belief, each fruit contains 613 seeds—which is the total number of religious precepts (commandments or *mitzvot*) in the Torah (the first five Books of the Judeo-Christian Bible). Pomegranates by the same tradition are a favorite food on the Jewish New Year, Rosh Hashanah. Depictions of pomegranates also featured on coins in ancient Judea, and even today it is not uncommon for a *sefer Torah* (the parchment scroll on which the Torah is inscribed, rolled between two wooden spools), when not being used, to have its upper handles encased in a pair of hollow sterling silver pomegranates, albeit stylized to the point of unrecognizability.

THE FRUCTIFEROUS ROSE FAMILY

The rose family Rosaceae contributes by far the largest proportion of edible fruit genera and species of all the plant world. By long tradition, the family was divided into four subfamilies—Rosoideae, Maloideae, Amydaloideae, and Spiraeoideae—but it has recently been convincingly proposed that the Spiraeoideae are the stock from which the Maloideae and Amygdaloideae were originally generated. In terms of fruits eaten by humans, this means that the overall family may (in somewhat simplified form) be considered thus:

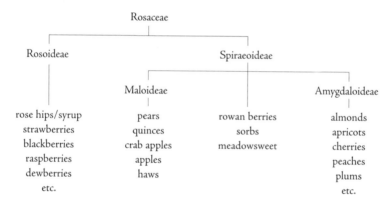

There are more than 1,000 individual species in the genus *Rubus* (which includes blackberries and raspberries) alone.

THE PEACH
Prunus persica

ROSACEAE

DESPITE THE SUPPOSED origin specified in the peach's Latin name (*Prunus persica*, in which *persica* means "of Persia"), peaches were actually first grown in China, where they have been cultivated from times of great antiquity. From China, peaches were taken to India and western Asia during the first millennium BCE, and were introduced to Europe at the behest of Alexander the Great (356–323 BCE)—who did indeed find them in Persia. Spanish explorers then took the peach to the Americas in the 16th century CE, before it had ever been seen in northern Europe.

The flesh of the peach (and of the very closely related nectarine) comes in either of two colors: yellow and white—although the white is a yellowish white. White peach flesh is generally sweet with little acidity, and is the preferred type in countries of the Far East. Yellow peach flesh generally has a perceptible tang of acidity, and is the preferred type in Europe and North America. There is another distinction between two other types of peach (and nectarine), depending on whether the flesh adheres to the central "stone" or not. Peaches in which the flesh adheres are described as "clingstones"; those in which the flesh does not adhere are called "freestones."

Fresh peaches do not keep well, but may be briefly stored at room temperature. Chilling in a refrigerator destroys some of the taste.

"Peach," color copper engraving by Pierre Joseph Redouté (1759–1840)

from Choix des Plus Belles Fleurs

RETOUCHED BY LANGLOIS

THE BOYSENBERRY

A boysenberry is a cross between the European red raspberry (*Rubus idaeus*), a blackberry (*R. fruticosus*), and a loganberry (*R. × loganobaccus*). It resembles an elongated blackberry and when ripe is a large (one-quarter ounce) compound fruit, with large seeds and a deep maroon color.

In the late 1920s George M. Darrow of the U.S. Department of Agriculture began tracking down reports of a large reddish-purple berry that had been grown on the northern California farm of a man named Rudolph Boysen. Darrow learned that Boysen had abandoned his growing experiments several years earlier and sold his farm. Unfazed by this, Darrow went to Boysen's old farm, on which he found several ancient vines surviving in a field choked with weeds. He took the vines to Walter Knott's farm in Buena Park, California, to be nurtured back to fruit-bearing health. Walter Knott was the first to commercially cultivate the berry in Southern California. He began selling the berries at his roadside farm stand ("Knott's Berry Farm") in 1932 and soon passers-by began coming back to buy the large succulent berries. When asked what they were called, Knott said "Boysenberries," after their originator. However, it was the other fruit preserves the farm stand sold that made it famous nationwide.

AVOIDING THE APRICOT

It is apparently a fervently held superstition among tank drivers in the U.S. Marine Corps that apricots are unlucky. They do not eat apricots, they won't allow apricots on board their tanks, and some of them refuse even to pronounce the word aloud. It all stems, they say, from a time during World War II when Marine Sherman tanks kept breaking down one after another for no evident reason—and the only common factor seemed to be the canned apricots in the soldiers' ration bags.

PEACHES OF IMMORTALITY

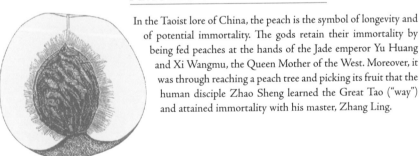

In the Taoist lore of China, the peach is the symbol of longevity and of potential immortality. The gods retain their immortality by being fed peaches at the hands of the Jade emperor Yu Huang and Xi Wangmu, the Queen Mother of the West. Moreover, it was through reaching a peach tree and picking its fruit that the human disciple Zhao Sheng learned the Great Tao ("way") and attained immortality with his master, Zhang Ling.

HOW TO GROW AN HEIRLOOM PEACH TREE

Choose somewhere very sunny with well-drained soil and, before planting out, fork in a lot of well-rotted compost to the site and perhaps make the soil more acidic by adding a little lime (peaches prefer a pH of 6.5); make sure that the hole in which the tree is to be planted is wide and deep enough. Buy plants as one- or two-year-old potted trees from a garden center or nursery; if the tree comes with a bag of peat around the roots, you may leave the bag there, but make slits in the sides to let the roots out in due course—avoid harming the roots. If your tree comes with the roots inside a burlap bag, remove it altogether. It's best to plant in winter to early spring.

1. Carefully drive a wooden stake vertically on one side of the hole in which the tree is to be planted.
2. Stand the tree in the hole parallel with the stake, spreading the roots in all directions; it should be planted to the depth that it was in its original pot.
3. Attach the tree to the stake, and fill in the hole with soil, rocking the tree very slightly with each spadeful to ensure the soil fills all the spaces between the roots and the stake.
4. Firm gently but often until the hole is filled; water in well, soaking it generously, and adding more soil if the surface level sinks. You will probably be able to remove the stake after the first year.
5. For the best results, use slow-release fruit-tree fertilizer spikes; otherwise, fertilize twice a year, using a high-nitrogen fertilizer (e.g. blood and bonemeal).
6. Every spring apply a dormant oil fruit-tree spray to kill as many insects as possible; in fact, a regular schedule of spraying against insects and plant diseases is essential for peaches, which are liable to die without protection. You may need to consider ways to minimize frost damage.

Fruiting: If large amounts of fruit appear to be forming, thin them out in late spring to ease the strain on the branches later; fruit for eating should be left on the tree until ripe (in summer and fall); authorities are divided over how much peaches really ripen, if at all, after being picked. It is likely that you will have to erect netting over ripening fruit to protect against bird (or animal) predation.

Pruning: The main pruning must be done between late winter and early spring, the intention being to remove branches to increase air circulation and sun penetration.

THE FIG

Ficus carica

MORACEAE

THE EDIBLE FIG may well have been the first plant ever cultivated by humans, possibly a thousand years before any cereal or vegetable crop. Carbonized evidence that strongly suggests fig cultivation dating from before 9000 BCE was discovered in a Neolithic village by the Jordan River in 2006. The fact that figs there had been grown on trees on which the fruit had not been pollinated (had been produced by parthenocarpic means) indicated the probability that they had not been growing wild. Fig trees, after all, are unusual in that when the fig wasp—its highly specialized, almost symbiotic, pollinator—is unavailable, they can nevertheless produce fruit from planted branches, and it is this method of fig production that most home gardeners take advantage of today. Commercial growers, on the other hand, may grow intermediate (or San Pedro) figs that require no pollination for the first (breva) crop, but do normally require pollination for the main crop. Or they may grow caducous (or Smyrna) figs that require pollination by the fig wasp together with the nearby presence of caprifigs (hermaphroditic common fig trees).

Figs are an excellent source of dietary fiber and calcium. It is the fiber that gives them their renowned laxative qualities. Dried figs contain even higher concentrations of fiber, calcium, and other beneficial minerals, and vitamin K. Most commercially grown fruit are dried or processed: fresh figs do not keep for long or travel well.

Figue violette.   *Ficus violacea*

P. J. Redouté . _ *38*

Bessin

"Figs," color copper engraving by Pierre Joseph Redouté (1759–1840)
from Choix des Plus Belles Fleurs
RETOUCHED BY LANGLOIS

HOUSEWARMING POMEGRANATES

When welcoming friends to a new home in Greece, a host may traditionally expect that his guests will bring a pomegranate as a symbol of prosperity, of good fortune, and of potential increase in the household itself.

THE COCONUT

In botanical terms, the coconut is not a nut but a form of drupe, of which the two outer layers—the mesocarp and exocarp—together make up the hard, hairy husk, while on the inside surface of the inner endocarp is the thick white albuminous layer of coconut flesh (or "meat").

Even if the coconut is not a nut (and English is one of very few languages that insist on referring to the tree *Cocos nucifera* as a coconut palm instead of, more correctly, a coco palm), each coconut does contain a seed from which, on germination, a root emerges through one of the three germination hollows present at the base of each fruit. It is because the tree prefers a maritime habitat, especially the beaches on tropical islands, that the main method in which the coconut disperses its seeds has evolved: coconuts are buoyant and highly water-resistant. Some have floated on ocean currents all the way from west Africa up to Norway and have remained viable.

Remarkably, because coconuts have for so long been so common floating in coastal waters around the Indonesian islands of Bali and North Sulawesi, a local species of octopus (*Amphioctopus marginatus*) has evolved to make specific use of coconut shells for shelter and for defense—the first invertebrate ever recorded (in 2009) as being able to use tools.

THE BIG BANANA

The banana "tree" is the largest known herbaceous flowering plant. It has a central "trunk" that is actually a thickish stem that grows upward from a corm but that can reach a height of twenty-five feet. The leaves are up to nine feet in length and two feet wide; in some regions of the world they are often used as emergency umbrellas.

MAPLE SYRUP

Yes, we all know—maple syrup is not a fruit. However, it is used as a fruit, and it comes from a tree—mostly *Acer saccharum*, the sugar maple. In fact, the syrup is made by repeatedly boiling the tree's sap (which is also the method by which maple sugar and maple taffy are produced). However, to get about one quart of maple syrup it is necessary to boil around ten gallons of sap.

TANGERINES, SATSUMAS & CLEMENTINES

What is the difference between these? Actually, not much. All three are in fact varieties of the mandarin orange (*Citrus reticulata*), which is itself a form of small, flattened (oblate) orange. Tangerines have a thickish, slightly redder peel and ripen mainly in midwinter. The satsuma is a thinner-peeled, seedless, sweeter variety that is the type most often contained in cans, but since 1990 has become almost as popular to eat "fresh." The clementine is also generally seedless, but is closer to a tangerine than to a satsuma. The tangerine is named after Tangier(s), the port in Morocco from which the fruit was imported to Europe. The satsuma is named after the province of Satsuma in Japan, although the fruit did not specifically originate there. (The townships called Satsuma in Alabama, Florida, Louisiana, and Texas were named after the fruit.) The name of the clementine is said to derive from the forename of the head of a church orphanage in Algeria, one Fr. Clément Rodier, often credited with growing the first in 1902.

WHY "THE BIG APPLE"

The Big Apple is, of course, New York, but although most people connect that name with the 1970s, that was in fact when it was repopularized (by a New York tourist board agency). The name was first made well known in the 1920s by the sports writer for the *New York Morning Telegraph*, John J. Fitz Gerald. Even then, he did not coin the term, which, in its earliest printed form, seems to have arisen in a comment by Edward Martin in his book *The Wayfarer in New York* (1909). What he said was that to anyone who lived and worked outside the place, New York appeared to be a city greedy for the best and the most of everything, including federal funding. However, Martin used figurative language to express this, stating that to outsiders it seemed that "the big apple gets a disproportionate share of the national sap."

SEA BUCKTHORN

Sea buckthorn (*Hippophae* spp.) is a low-growing coastal shrub that thrives on seashores all around Asia and Europe. Female plants produce small orange-colored berries that are soft and juicy, but are too acidic to eat raw: they can, however, be baked in pies or turned into preserves. The fruit has a high vitamin C content—by weight about fifteen times greater than that of oranges—and partly because of that, but also because the fruit contains useful amino acids, minerals, and vitamin E, extracts are used in nutritional supplements for infants in some Nordic countries. Skin creams and liniments can also be processed from the juice.

THE BLACKBERRY
Rubus fruticosus

ROSACEAE

THERE ARE QUITE a number of different species of the genus *Rubus* that produce blackberries, although *Rubus fruticosus*, the common blackberry, is indeed by far the most common, and is the species generally found growing wild. Blackberries can propagate themselves by the form of asexual reproduction called apomixis (basically, a kind of cloning that results in offspring that are genetically identical with the parent plant, thus over two or three generations equating to a separate microspecies—of which several hundred have by now been officially recorded). At the same time, the commercial growing of blackberries has increased dramatically worldwide since the 1980s, many growers concentrating on producing *Rubus fruticosus* cultivars with particularly advantageous qualities and properties. One of the outcomes of this has been the creation of a third type of blackberry plant in addition to the older trailing and erect types: the semierect thornless blackberry.

Oregon has, since the turn of the millennium, led the world in the volume of cultivated blackberries produced—averaging more than forty million pounds (20,000 tons) per year, especially featuring the cultivar 'Marion'—although Mexico is fast catching up, and in Europe, Serbia is not that far behind either. Wild blackberries are decreasing in importance, although by 2010 the overall annual consumption of wild blackberries was still estimated at more than 11,000 tons worldwide.

"Brambleberry," color lithograph by E. Walther (n.d.)
from Die Pflanzen-welt
J. F. SCHREIBER, ESSLINGEN (1891)

GIN & SLOE GIN

Gin is made by redistilling pure grain alcohol together with the fully grown but immature green "berries" of the coniferous tree *Juniperus communis*, the common juniper, and a much smaller quantity of other herbs and plant extracts (including citrus zest). Juniper berries are, however, false berries in that they are actually female seed cones in which the scales have merged and become fleshy, so that when mature they look like very small, hard, dark blue-purple plums. Along with pine nuts and some resinous products, juniper "berries" are, thus, examples of rare food ingredients derived from a coniferous tree. A number of other juniper species also bear "berries," but most are too acidic for consumption or for flavoring, and one or two are toxic to humans.

In northern Europe, during the Middle Ages, it was thought that the berries of the common juniper had considerable medicinal properties, and gin was one preparation used—tragically, with little success other than perhaps anesthetic—to try to ward off the bubonic plague. Because it was then made primarily in the Netherlands, gin was also called "Hollands" and might be said to evoke "Dutch courage." Sloe gin is what is called a compound gin—an infusion of the fruit called sloes with a little sugar in gin, so forming a sloe-flavored version of juniper-berry gin—in which no additional distillation is involved. Sloes, which individually look remarkably like juniper berries, are small, dark, and relatively acidic members of the plum family, growing on a woody shrub otherwise called the blackthorn (*Prunus spinosa*), from straight stems of which canes (and in Ireland the clubs known as shillelaghs) are made.

THE LOQUAT

Formerly classified as a type of medlar (*Mespilus* spp.) and occasionally still described as the Japanese medlar, the loquat (*Eriobotrya japonica*) is, in fact, more closely—but even then fairly distantly—related to the apple. What is most unusual about the loquat is that the tree produces its headily aromatic flowers in fall or early winter and the golf-ball-sized round or oval orange-colored fruit appear in late winter or early spring. Inside the smooth or sometimes lightly furry skin is juicy flesh that is sweetish but with a tang to it, surrounding up to five large brown seeds. The skin is usually peeled off to eat the fruit fresh; the seeds should not be chewed or swallowed. Otherwise, loquats may be poached in their own syrup, or made into preserves and jellies, or included in chutneys. It is unwise to eat a large quantity of fresh loquats because the flesh contains substances that have a mild but cumulative sedative effect that may last for up to a day.

HOW TO GROW HEIRLOOM BLACKBERRIES

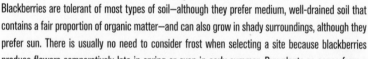

Blackberries are tolerant of most types of soil—although they prefer medium, well-drained soil that contains a fair proportion of organic matter—and can also grow in shady surroundings, although they prefer sun. There is usually no need to consider frost when selecting a site because blackberries produce flowers comparatively late in spring or even in early summer. Buy plants as canes from a garden center or nursery. Planting in late fall is best, but any time up to early spring unless the soil is frozen or waterlogged. Because canes can last for up to at least fourteen years, there should be no need at this early stage to consider propagation for future harvests.

1. Around sixteen days before planting, dig in a good quantity of organic matter (compost) to enable the soil to retain moisture.
2. Follow the retailer's advice on how far apart to site the canes; vigorous-growing varieties require more space (some as much as thirteen feet). Plant to a depth that keeps the crown of the root level with the surface of the soil; generally, this involves digging a hole five inches deep. Spread the roots out and cover in soil, firming the soil down using your knuckles; cut the top of each cane above the soil surface down to about ten inches and water well, but be careful never to waterlog.
3. The canes may or may not need support; blackberries require air circulation, and you may find that training some (but not all) along wires attached to posts assists this.
4. Dress the canes annually with well-rotted compost or use a long-lasting fertilizer (such as bone-meal). There is no need to consider pollination: blackberries are self-fertilizing.

Fruiting: From the second year—usually the year in which fruit first form—harvest the fruit a little at a time and frequently, and not all at once; this encourages the formation of more fruit. Pick the fruit in dry weather to prevent picked fruit from rotting, and in any case eat fruit within a day of picking.

Pruning: As soon as the year's crop of blackberries is over, cut the stems that have produced fruit during the year back to ground level; do not touch stems that have not produced fruit during the year—blackberries have a biennial cycle, and these stems will produce fruit next year.

THE BANANA
Musa acuminata

MUSACEAE

IT WAS ONLY in the 1950s that it was realized that the many species of banana then listed under separate scientific names were in fact hybrids and cultivars of a much smaller number of genuine species. The redefinition process is still ongoing, however, because far from simplifying matters, sorting out the original parent species of the hybridized cultivars has been fiendishly complicated by the fact that the hybridization process evidently also involved "parents" that were themselves hybrids. This meant that account had to be taken not just of the genetic contribution of species, but of the comparative proportions of those contributions in the makeup of individual cultivars. It is now possible to say with some assurance, though, that most of the bananas (and plantains) eaten worldwide today contain genetic elements of *Musa acuminata*, and some contain a genetic contribution additionally from *M. balbisiana*. It is because of this long history of hybridization—mostly as a result of deliberate intervention by humans in Southeast Asia, especially Papua New Guinea—that cultivated bananas today are virtually seedless: the seeds appear as insignificant black specks within the flesh of the fruit. The few remaining types of wild banana correspondingly have a lot of hard seeds.

By volume, three-quarters of the banana's flesh is water. The rest is mostly fibrous matter that is also surrounded by fibrous phloem bundles (the "strings") that connect with the skin. The skin can also be eaten raw or cooked.

Musa acuminata

"Bananas," color lithograph after drawing by Richard Bridgens (1785–1846)

from Aus: West India Scenery

R. JENNINGS & CO., LONDON (1836)

COCONUT WATER

The watery white "juice" inside a coconut contains sugar, fiber, proteins, vitamins, minerals, and antioxidants, and is accordingly often used in isotonic sports drinks to contribute to the electrolyte balance in the body. However, until the coconut is opened, "coconut water" is also sterile, and it has the additionally useful medical property of mixing well with human blood—for which reasons it can be used in emergency blood transfusions (as it sometimes was during World War II).

JACKFRUIT

The jackfruit (*Artocarpus heterophyllus*) is a close relative of the breadfruit (*A. altilis*) and can be described as a giant version of it. In fact, the jackfruit—a native of southern and Southeast Asia, but found also in coastal East Africa—is the largest of all tree-borne fruit. Some specimens measure up to three feet long and twenty inches wide, and can weigh as much as eighty pounds. Perhaps appropriately to its outsize dimensions, the jackfruit has probably more completely unrelated names given to it by different communities around the world than any other fruit.

The fruit has a history of cultivation in India stretching back at least six thousand years. Yet in the Tijuca Forest National Park, in Rio de Janeiro, Brazil—in which jackfruit were planted at its inauguration during the 19th century—no fewer than 55,600 jackfruit saplings were cut down between 2002 and 2007 to prevent them from invasively dominating land set aside for native tree species.

RIPENING PEARS

Pears ripen from the inside out. To hasten the ripening of a pear once picked, keep it at room temperature and place it next to one or more bananas in a bowl. To slow the ripening of a pear, put it in the refrigerator. Either way, because pears ripen from the inside out it is possible to tell fairly simply whether a pear has reached ripeness by applying gentle thumb pressure to the neck, top, or stalk end of the pear to check for its firmness or its yielding. If it yields to this gentle pressure, it should be ripe, sweet, and juicy.

PICKLED PLUMS

Pickled plums are readily available in much of Asia, and especially Japan, where a special type of plum is used, along with a considerable amount of salt. Pickled plums are often used in sandwiches or as a cold side-dish to accompany a hot meal.

GOJI BERRIES

Available in many health-food stores in North America and Europe, goji berries are the dried fruits of two species of the boxthorn—*Lycium barbarum* and *L. chinense*. For at least 300 years both the growing plants and fresh fruit have been known as wolfberries. The *Lycium* genus belongs to the family Solanaceae, to which the potato and tomato also belong. On the tree, wolfberries hang from branches in stalked clusters of three to six, like little red tubes with rounded ends, up to four-fifths of an inch long and about half of an inch wide. Each berry contains between ten and sixty tiny yellow seeds.

The harvested fruit are not found in fresh form outside their area of growth. Virtually all are dried immediately after harvesting in a careful way to avoid bruising the berries, although the extent of the drying process may differ between areas. The dried fruit can be cooked or boiled and used in soups or as flavorings in meat dishes with other herbs.

However, since 2000, there has been a considerable effort on the part of commercial growers in China, Tibet, and some Himalayan regions to publicize the medicinal properties of "goji berries," notably as sources of antioxidants (which are beneficial for inflammations and cardiovascular disorders) and protection against the infirmities of old age (including some forms of cancer, and protein- and vitamin-deficiency disorders). Some of these claims have been received with skepticism—and one or two with outright rejection and consequent official warnings—by scientific bodies and governmental authorities around the world.

ICY ORANGES

Orange trees are subject to damage when the ambient temperature drops well below freezing. Commercial growers in countries where such temperatures may occur in winter must be prepared to take precautions. As improbable as it sounds, one way to protect the trees when severe frost is forecast is to spray the trees with water. As long as the water on the trees is turning to ice on the branches, the trees themselves will remain at or around thirty-two degrees Fahrenheit and will drop no lower, even if the surrounding air temperature falls many degrees below.

PEACH ALLERGY

An allergy to peaches is a relatively common form of hypersensitivity, caused most often by the presence of specific proteins in the skins of fresh peaches—the reaction is far less common with peeled, boiled, or canned peaches. Symptoms range from the severe (anaphylaxis) to the very unpleasant (such as hives) to somewhat milder but upsettingly uncomfortable respiratory and gastrointestinal disorders.

THE POMEGRANATE

Punica granatum

LYTHRACEAE

BECAUSE OF ITS MULTIPLICITY of seeds, the pomegranate has from ancient times been a symbol of fertility—every single seed, after all, might in the future produce many more fruit. The ancient Greeks certainly thought of it in this way, with particular reference to their harvests of cereal crops. This is why it was a pomegranate that the mythical corn-maiden Persephone ate while in the underworld, so causing her to have to spend four months (the months of winter) below the surface of the earth every year before reappearing in spring as the shoots of the next grain harvest. To the Romans, however, the connotations of the pomegranate's symbolism were less agricultural and more social: it represented fertility in marriage. A bride tended to wear a wreath of pomegranate twigs and leaves. That kind of symbolism is still observable even today in such countries as Armenia and Greece.

Sadly for the idea of fertility in marriage, the pomegranate was also an emblem on the coat of arms of Queen Catherine of Aragon (1485–1536), first wife of King Henry VIII of England (ruled 1508–47). When Henry and Catherine could not produce a male heir, the king put her aside and in 1533 married Catherine's maid of honor, Anne Boleyn. As queen, Anne's first royal ordinance was to have a special badge designed for herself and her staff— featuring a crowned and sceptered white falcon standing on a pomegranate.

Grenade.   *Grenadier punica.*

P. J. Redouté. _ 50.   Victor

"Pomegranate," color copper engraving by Pierre Joseph Redouté (1759–1840)

from Choix des Plus Belles Fleurs

RETOUCHED BY LANGLOIS

THE PASSION FRUIT
Passiflora edulis

PASSIFLORACEAE

THERE ARE TWO very different types of passion fruit, and exactly how they are related remains unclear even after genetic investigation. However, they are related, and apparently closely. The two forms are the purple passion fruit (*Passiflora edulis*), which is cultivated mainly in Africa and India and may alternately be called the granadilla or grenadilla; and the yellow passion fruit (often for convenience classified as *Passiflora edulis* var. *flavicarpa*), which is cultivated mainly in northern and northeastern South America and may alternately be called the golden passion fruit or maracuya. The purple passion fruit is the size and shape of a large lime, whereas the yellow passion fruit is the size and shape of a grapefruit. The skin of both types is fairly light, and encloses a firm to soft juicy interior packed with seeds. Passion fruit is mostly eaten fresh or grown for its juice, which has many culinary applications.

The passion fruit was first so called by Catholic missionaries in South America, who used various visual elements that they saw in the fruit to teach the indigenous peoples about the Passion of Christ. The corona threads of the passion flower might represent the crown of thorns; the vine's tendrils might be the scourges; the three stigmas might be the nails in hands and feet; the five anthers might be the five wounds; and the ten petals and sepals might correspond to the ten apostles (twelve less Judas, who betrayed Christ, and Peter, who had denied Christ).

GRANADILLA, *Pentaphylles flore cœrulæ magno. Beech Ind. tab. 11. 81.*

J. Van. Huysum.

"Passion Flower," illustration by Jacobus van Huysum (1687–1740)
Private Collection

FRUIT COCKTAIL

In most countries of the world, "fruit cocktail" is just an alternative expression for "fruit salad" and refers to a dessert (or less commonly, an appetizer) that comprises fresh or canned fruit—diced, sliced, or cut—in juice or syrup, with or without cream or sauce. If fresh fruit are used, the ingredients may vary considerably, depending on where in the world the cocktail is being made, the season of the year, and the creativity of the maker. Around the world, canned fruit cocktails likewise vary in ingredients, although to a lesser extent, depending on locality and resources … Except in the United States, where the U.S. Department of Agriculture has formally decreed that canned "fruit cocktail" should comprise specific fruits in specific (although relative) proportions:

FRUIT	PROPORTIONS (%)	ADDITIONAL NOTES
peaches – diced	30–50	*yellow varieties only*
pears – diced	25–45	*any variety*
grapes – whole	6–20	*seedless varieties only*
pineapple – diced	6–16	*any variety*
cherries – halved	0 to "a little"	*any light sweet or artificial red variety*

CARAVAGGIO'S BASKET OF FRUIT

Hailed as a masterpiece, and as significant to the world of still-life art thereafter, the painting known as *Basket of Fruit* by Michelangelo da Caravaggio (1571–1610) was completed in about 1599 and now hangs in the Biblioteca Ambrosiana in Milan, Italy. It depicts a realistically vivid selection of choice summer fruit. However, not everything about the fruit is what it ought to be. In fact, although certainly not obvious at first glance, most of the fruit, and the leaves and twigs that accompany them, have been damaged by insect predation, worm penetration, or fungal growth, and a few of the fruit themselves are overripe and decaying. It is not known what the point of this is—or even if there is any point—but if not, that is hard to believe, for Caravaggio was notorious for including "coded messages" in his paintings. Various possible explanations have been suggested, but perhaps we will never know what lies behind it in this case.

MAGIC MULBERRIES

Unripe mulberries—and the green leaves and shoots of the mulberry plant—contain a white resinous fluid (latex) that, when swallowed, is liable to cause gastric upset; however, it can also result in a heightening of the senses and in intoxication that may thereafter lead to hallucinations.

HOW TO GROW AN HEIRLOOM AVOCADO

Avocados are native to Mexico southward to northern South America, and, unless you live in a tropical or subtropical region, they are best grown indoors in a tub or large container. However, be aware that it may be ten years before your avocado tree eventually fruits, and under the right conditions it could in due course grow to sixty-nine feet high.

1. There are three ways to grow from a pit: stick three toothpicks tripodwise into the cleaned pit and suspend it partly in water within a glass, pointed side up, so that the base of the pit is submerged to a depth of half to one inch; expose the cleaned pit to strong and direct sunlight until it starts to split; or plant the pit completely covered in a pot of potting mix, place it in dark and humid surroundings, water well, check every week to see if a shoot is appearing, and when it does, remove it from its surroundings.

   In any of these cases, the roots should develop and a shoot should rise vertically from a split in the pit; when this happens, expose the plant to as much hot sunlight as possible and plant it in a pot. Use a pot with a diameter of around four-and-a-half inches, filling it with enriched, peat-free potting mix and making sure of good drainage.

2. When your plant reaches a height of one foot, cut it down by half so it spreads outward, not upward.

3. When the roots fill the available space, transfer the plant to the largest container you have that is compatible with the temperature requirement.

4. Each time the plant grows another six inches, pinch off the latest sets of leaves on top.

5. If and when the plant flowers, it will require pollination in order to fruit; if in a greenhouse or sun room, leave a door open for bees to enter; otherwise (if feasible), move the plant outside only on hot, sunny days and bring it back in again each time afterward.

Harvesting: Eventually, if you are lucky, numerous fruit will appear on your tree. Do not wait for them to ripen before picking: pick when they are mature and have reached a suitable size, and then let them ripen off the tree. Once picked, avocados ripen in a few days at room temperature, and they ripen faster if stored near other fruits such as apples or bananas (because of the influence of ethylene gas).

THE RED CURRANT
Ribes rubrum

GROSSULARIACEAE

THE RED CURRANT (*Ribes rubrum*) is a member of the gooseberry family Grossulariaceae, native to northern and western Europe. Its translucent red berries grow in large numbers on the deciduous shrub—indeed, an established bush can produce up to around nine pounds of fruit between mid- and late summer—and have a pleasantly sweet flavor with a touch of acidity. The white currant, which is often listed as a cultivar of *R. rubrum* and occasionally described as a different species altogether, is, in fact, no more than a recurring albino form of *R. rubrum*. It has a slightly sweeter taste. Both forms may be eaten raw, baked in pies or tarts, or turned into preserves or jelly (the latter particularly as a condiment with roast lamb).

Red currants and their close relatives black currants (*R. nigrum*) were until the early 20th century popular fruit in the United States. Then, along with gooseberries, they were identified as potential host plants to the vicious fungal disease white-pine blister rust, and correspondingly a danger to the lucrative white-pine logging trade, so their cultivation was immediately banned. As new information came to light, the federal ban was redesignated to the authority of individual states in 1966, and by 2005 several states (including New York, Connecticut, Vermont, and Oregon) had not only lifted the ban, but were encouraging cultivation. The ban nonetheless remains in force in some states.

Groseiller rouge.   *Ribes rubrum.*

P. J. Redouté. 51.   Langlois.

"Red Currant," color copper engraving by Pierre Joseph Redouté (1759–1840)
from Choix des Plus Belles Fleurs
RETOUCHED BY LANGLOIS

WHEN IS A BERRY NOT A BERRY

When it's a strawberry. Strawberries are of course not really berries at all. Although they are of the same overall family, Rosaceae, as most of the ordinary berries, they are, in fact, a very specialized form of fruit like no other, sometimes known as a false or secondary fruit, but otherwise described as a "seed organ." The "flesh" of the strawberry is not derived from the plant's ovaries, but from the peg-like stock around and under which the rounded bowl-shaped hypanthium grows, which has the seeds on the outside. The average strawberry has 200 seeds.

THE CULINARY APRICOT

As an accompaniment to cooked dishes, apricots are extraordinarily useful. Apricot halves go particularly well with fried ham or cooked in chicken casseroles. The combination of steamed asparagus and sliced apricot is hailed by many as utterly delicious with steak, especially with a sprinkling of ginger and/or cinnamon on top.

APPLE TREES & ROYALTY

Appeldoorn is a reasonably large city (population 136,200 in 2008) in the province of Gelderland near the center of Holland, about sixty miles southeast of Amsterdam. The city was apparently founded in the eighth century as a trading settlement at a major crossroads—a crossroads evidently marked by an apple tree, which is what the original form of the city's name (Appoldro) meant. Within the boundaries of the city lies the grand palace of Het Loo, now a national museum incorporating the Royal Library and state military archives, but once a favored country seat of the Dukes of Gelderland, thereafter to become the Dutch royal family. It was William III of that family who, when he and his wife Mary became joint sovereigns of England in 1689, undertook the enlarging and refurbishing of the palace of Het Loo to the magnificent form in which it can be seen today. Slightly ironic, however, is that in the city named after an apple tree stands the favorite residence in Holland of members of the royal Orange family.

LITCHIS AS PACIFIERS

In China it is an age-old tradition to use a peeled litchi as a pacifier for babies from the age of about nineteen weeks. It not only keeps them quiet, but adds valuable natural nutrients to the infants' dietary intake.

FIGS & THE ROMANS

It is often said that the ancient Romans were very fond of figs—that Cato the Elder (second century BCE) wrote about the different varieties that were available in the Italian markets of the time, that Cleopatra regarded figs as her favorite fruit of all, and that images of bowls of figs were to be seen on the walls of Roman villas. And, of course, that may all be true and rightly be taken at face value. However, figs were sacred to Bacchus—and Bacchus was the god in whose name the festivals known as the Bacchanalia were undertaken, originally attended only by women, and on only two days a year, yet later opened also to men and held five times a month, with a reputation for extreme lewdness. By 186 BCE, so notoriously licentious had the Bacchanalia become that the Senate formally investigated, and finally banned them.

Perhaps because of this background, the word "fig" in Latin took on crude sexual connotations, which remained in the language and in Latin-derived languages for hundreds of years (as quoted from Spanish, for example, by Shakespeare). So whereas Cato the Elder was a genuinely ex-plebeian farmer who attempted to revitalize rural agriculture and, therefore, discussed figs for the fruit they were, it may be that to describe Cleopatra as liking figs was intended to imply that she was prepared to bestow favors on any senior Roman official who came to call; and to describe images of figs on the interior walls of villas may beg questions: what kind of houses were they, with what kind of residents?

THE "ORANGE BLOSSOM SPECIAL"

The well-loved tune "Orange Blossom Special" was written for the fiddle in 1939 by the brothers Ervin and Gordon Rouse of Jacksonville, Florida. It was named after a deluxe ("all-Pullman") railroad train service between New York and Miami (and, in winter, Tampa and St. Petersburg) that operated from 1925 as a combined service of three separate railroad companies in sequence over the whole distance, the longest proportion by the Seaboard Air Line Railroad. The "Orange Blossom Special" was named and advertised as a luxury rail service to the orangeries of Florida (of which the orange blossom is the state flower). The train's last run was in 1953, by which time the tune had become so well known at bluegrass music festivals that for at least a decade no bluegrass fiddle-player would be engaged for any band performance unless he could play it proficiently.

THE PLUM

Prunus domestica

ROSACEAE

THE ORDINARY RED-PURPLE European garden plum is of the same genus (*Prunus*) as peaches and cherries, but is of the same subgenus (also called *Prunus*) as, for instance, greengages (which are green), damsons (which are blue-purple), and mirabelle plums (which are golden). Most of the members of the subgenus may be easily hybridized, so there are in addition many intermediate types of plum. However, even the ordinary garden plum comes in a huge number of different varieties, the edible but somewhat acidic-tasting skins of some being bluer or redder than the standard red-purple, and the flesh of others being red, green, yellow, or even white.

Dried plums have been known as prunes—although, technically, "prune" is also the name of any of the more than a thousand plum cultivars that are bluer and more egg-shaped than the standard roundish plum, and that may also be described as "freestone" (the flesh comes away from the central stone without adhering to it). Fresh or dried, plums are famous for their laxative effect on the human digestive system, which is due largely to the presence in the fruit of plenty of dietary fiber, the glucose-derived carbohydrate sorbitol, and the aromatic plant substance isatin. So well known is this effect that the sellers of prunes for nonmedical purposes (e.g. for eating cooked as a dessert) have over the last two decades insisted on calling them "dried plums" instead.

Prune Royale.   *Prunus Domestica.*

P. J. Redouté. _114.   Langlois

"Plum," color copper engraving by Pierre Joseph Redouté (1759–1840)

from Choix des Plus Belles Fleurs

RETOUCHED BY LANGLOIS

HACKBERRIES

Hackberries (*Celtis* spp.) are not well known despite being widespread in warm temperate regions right around the Northern Hemisphere. The tree was recently reclassified as being in the family Cannabaceae (the hemp family), having formerly been included in the elm family (Ulmaceae) or in a family all of its own. Its fruit are tiny round drupes that ripen from green to yellow. They contain little juice, but are sugary-sweet (much like a date). The major European species is *Celtis australis*, also known as the European nettletree; major Asian species are *C. sinensis*, the Chinese hackberry or Chinese nettletree; and *C. jessoensis*, the Japanese hackberry. Perhaps the most impressive property of the hackberry tree is that it tolerates bending, without breaking, more than any other type of wood.

THE CASHEW APPLE

The true fruit of the cashew tree (*Anacardium occidentale* or *A. curatellifolium*) is the cashew nut. However, what looks as if it is the fruit (but it is technically the "receptacle" of the flower) grows immediately above the nut and, when the nut in its double shell is near ripening, it fills out to become plump and fleshy, in shape much like a red or yellow pepper. Inside the waxy skin of this cashew apple—which is thus a false fruit, or pseudocarp, and is hardly comparable with a real apple, either—is a fibrously spongy yellow pulp with a quantity of juice that has a sweetish but tangy taste.

As the skin of the fruit bruises easily, and because the cashew *nut* is of far greater commercial value as a crop, cashew apples are rarely harvested for transportation. Instead, they are eaten fresh or mashed for their juice, close to the location of their growth. The juice, which is rich in vitamin C, is highly popular as a cold drink in the cashew apple's native Brazil, and is fermented to become an alcoholic beverage in India and East Africa.

APPLES ACROSS EUROPE

In 2006 the European Union Commission published a report on tests carried out to detect residual pesticides in fruit for sale in member states (both domestic and imported fruit). In the report it was noted that no fewer than sixteen percent of all apples sold as fresh fruit across all EU member states (a statistic that therefore does not apply in individual countries) were contaminated with trace amounts of the organophosphate insecticide chlorpyrifos, a substance that is officially categorized as "moderately hazardous."

HOW TO GROW AN HEIRLOOM PLUM TREE

These instructions are for growing a plum tree purchased from a garden center or nursery as a two-year-old potted cordon, for planting in open ground with a stake support. This is the simplest method, but there are a few other forms of tree (such as the fan, the pyramid, and the bush) for which the details of growing care may be similar, but for which considerably more care is required in preparation and attention to support (particularly a wall). Of course, plum trees may be grown in tubs.

Select a site that is sunny, with well-drained soil; plums are tolerant of most soil types. It is best to plant in late fall to early spring.

1. Fork a general fertilizer into the soil—do not use fresh manure (which may damage the roots)—and make sure that the hole in which it is to be planted is wide and deep enough to accommodate the roots in both dimensions.
2. Carefully drive a wooden stake vertically on one side of the tree-planting hole. Stand the tree in the hole parallel with the stake, spreading the roots in all directions; it should be planted to the depth that it was in its original pot—in any event, do not bury the knobbly junction between the rootstock and scion. Attach the tree to the stake, and fill in the hole with soil, rocking the tree very slightly with each spadeful so as to ensure the soil fills all the space between the roots and stake; firm gently but often until the hole is filled. Water in well, but do not waterlog.
3. Apply a mulch annually in spring of well-rotted compost or manure; a dressing of nitrogen-rich inorganic fertilizer is recommended in late winter.

Fruiting: If large amounts of fruit appear to be forming, thin them out in early summer to ease the strain on the branches later; leave fruit for eating on the tree until ripe (in late summer and fall); harvest fruit for preserving earlier. Consider whether you need to protect from frost or bird predation.

Pruning: Pruning is essential and must be done between spring and midsummer; moreover, pruning must be severe for the first three years, no more than one quarter of an inch above the buds; slope all pruning cuts away from the buds (to avoid water contamination).

THE BREADFRUIT
Artocarpus altilis

MORACEAE

THE BREADFRUIT, growing on the breadfruit tree *Artocarpus altilis* (in which *Artocarpus* is a direct translation of "breadfruit"), looks like a slightly elongated green grapefruit with a stippled outer surface. Inside, however, it is nothing like a hesperidium citrus fruit, and, indeed, it is not really a fruit at all because it corresponds to an enormously complex development of the perianth comprising between 1,500 and 2,000 flowers.

The great thing about the breadfruit is that one tree produces a huge quantity of fruit, and in more than one burst per year. From the South Pacific to southern India, the number of fruit harvested per tree averages between 150 and 200 a year, which is why since ancient times they have been propagated in the form of cuttings and air-layered plantlings by seafarers migrating from island to island. The breadfruit, with its high starch content, even today accordingly remains the staple diet of many tropical-island residents. However, it is not eaten raw. It is boiled, baked, fried or broiled, after which the taste has been compared with that of freshly made bread or perhaps potato. It is no wonder that South Pacific myths about the origin of the breadfruit almost all center on the fruit being the result of divine intervention which in the nick of time saved humans from extinction by starvation.

"Breadfruit Tree," illustration by unknown artist (n.d.)

Private Collection

BOTANY LIBRARY, THE NATURAL HISTORY MUSEUM, LONDON

HOW SWEET THE ORANGE IS

The sweetness of orange juice (and also of other liquids that contain sugar, including fermenting wines) is measured in degrees Brix; a high Brix number means a high sugar content. In 2009 the standard sugar content of oranges from Florida was quoted as forty-two Brix, whereas oranges from Brazil contain enough sugar to measure sixty-five Brix.

THE FRUCTIFEROUS SAPINDALES ORDER

The botanical order of flowering plants known scientifically as the Sapindales includes three major families of trees that bear fruit consumed by humans. It also includes such members as horse chestnuts, the hardwood mahogany, and the trees from which the resinous substances frankincense and myrrh are derived.

The three fruit-bearing families are:

Rutaceae	Which include all the citrus fruit, but also include the noncitrus limeberry (*Triphasia trifolia*) and the bael (*Aegle marmelos*, also called the Bengal quince); various herbal plants, such as rue (*Ruta* spp.) also belong to this family.
Anacardiaceae	Mangoes (*Mangifera* spp.) and cashews (*Anacardium occidentale*; the cashew nut grows suspended beneath the "cashew apple" (see page 64), which is itself an edible false fruit, or pseudocarp).
Sapindaceae	Litchis (*Litchi chinensis*, almost always eaten fresh because, when kept or canned, the delicate perfume is lost); rambutans (*Nephelium lappaceum*, a kind of red spiky-haired litchi); the longan (*Dimocarpus longan*, in which *longan* is adapted Chinese for "dragon's eye," a rounder, sweeter, smoother litchi); and the mamoncillo or canepa (*Melicoccus bijugatus*, an ovoid green drupe, also called limoncillo, with a flavor reputedly between lime and litchi that ranges from sour to sweet).

There are also the Aceraceae, which some claim are within the family Sapindaceae, too. These include the maples (*Acer* spp.), which, although they do not bear edible fruit, do in some cases at least largely contribute to a much-loved edible product, maple syrup, made from the sap of the sugar maple (*A. saccharum*, whose leaf in its fall-red form appears on the national flag of Canada) or the black maple (*A. nigrum*).

THE PAPAYA & BIRTH CONTROL

It has long been a popular myth that the flesh and seeds of unripe papaya have properties that promote contraception and abortion in women and infertility in men. Women slaves on the plantations in the West Indies used to eat green papaya to try to avoid becoming pregnant.

When papayas began to be grown commercially in India, Sri Lanka, Pakistan, and Bangladesh, the unripe fruit quickly became a widespread folk "remedy" in those countries for the same purpose. Medical science has since confirmed that there is a good deal of truth in this "myth," and that the effects are increased, the more the unripe papaya is consumed. However, ripe papaya has little or no such effect, but nonetheless should be avoided by women who want to become or to remain pregnant, just to be on the safe side.

THE WORLD'S LARGEST CHERRY PIE

The prestige in producing the World's Largest Cherry Pie is more significant than you might imagine, although it has to be said that to date "largest" has generally meant "heaviest." In 1976 a town called Charlevoix close to the heart of the cherry-growing area on the northeastern shore of Lake Michigan, as part of its annual Cherry Festival—and using a specially constructed baking pan and a specially built oven—produced a pie weighing 17,420 pounds. Eleven years later (in 1987) the town of Traverse City, some fifty miles south, and by then even more famous for its own annual Cherry Festival, thrashed its local rival by producing a pie in the same fashion weighing no fewer than 28,350 pounds and with a diameter of seventeen-and-a-half feet—and had it formally recorded in *The Guinness Book of Records* of the following year (1988).

Unhappily for Traverse City, national commercial fruit production interests then moved in, and the impetus for local cherry-oriented eminence was lost, with the result that the comparatively tiny settlement of Oliver, British Columbia, Canada (about 160 miles east of Vancouver, and maybe 1,600 miles west of Lake Michigan), in turn, and apparently in like manner, beat all previous records by producing a cherry pie only five years later in 1992 that weighed 39,683 pounds.

THE NECTARINE

Prunus persica

⚜

ROSACEAE

THE NECTARINE is in almost every respect identical to a peach. It is scientifically classified as the same species, and has only one real difference— and a very obvious difference, at that—its skin. Whereas the peach has what is called "skin fuzz" (perhaps somewhat better described as "a velvety texture"), the skin of the nectarine is smooth. The difference is caused by the presence of dominant or recessive genes. The genes for a peach with "skin fuzz" are dominant; those for the smooth skin are recessive, and nectarines, therefore, only occur as the result of a combination of two or more recessive genes in the absence of dominant genes. They are not mutants; they are merely "sports." They can sometimes appear quite naturally among peaches on peach trees. However, most are, of course, commercially grown as nectarines these days, and those who sell them prefer to suggest that they are a fruit quite independent from peaches.

Like peaches, nectarines can have "white" or yellow flesh and be described as "clingstones" or "freestones." Yet the average nectarine is slightly smaller than the average peach, and its skin may look a little redder than a peach's (if only because of the lack of "skin fuzz"—a lack that actually makes the skin itself more delicate and potentially more subject to bruising).

Pêcher à fruits lisses.

"Nectarine," color copper engraving by Pierre Joseph Redouté (1759–1840)
from Choix des Plus Belles Fleurs
RETOUCHED BY LANGLOIS

A BANANA REPUBLIC

The term "banana republic" has long been used as a pejorative term by capitalist countries, especially at one time or another in relation to almost all countries in Central America. To be strictly accurate, only Costa Rica, Honduras, and Panama have ever been nation states in which the economy has been truly dominated by the banana trade.

THE FRUIT MACHINE

A fruit machine to most people is a one-armed bandit in which one places a coin, pulls the lever, and hopes that three of the same fruits line up in the center of the dial, releasing a shower of coins or tokens. However, another was an ill-conceived mechanical device, also known as the "fruit machine," and it was supposed to detect homosexuals in the Canadian Civil Service between 1950 and 1973. It was intended to detect minor but revealing bodily reactions to sensuous pictures and words with which Mounties were confronted during a lengthy test program. It is not known how many homosexual and heterosexual men were deemed to have failed the test, although the penalty for failing at that time was immediate arrest. According to reports, it was only after ten years of formally official use that the machine was consigned to the scrap heap as a monumental failure.

HOW TO WEAR A WATERMELON

The traditional helmet color of
the Saskatchewan Roughriders football team of
the Canadian Football League is dark green. However, over
the last decade or so, fans following their team have taken to buying
watermelons in local markets, hollowing them out, and wearing them as
"helmets" to support their heroes from the grandstands. Before the Grey Cup
championship final match in Calgary, Alberta, in November 2009 (between
the Roughriders and the Montreal Alouettes), thousands of fresh but out-of-
season watermelons had accordingly to be imported into Calgary grocery
stores, not only so that fans would be able to attire themselves in their
preferred headgear, but also so that local residents could join in the
celebrations in an appropriate manner if the Roughriders
won. Despite the watermelons, the Roughriders
were narrowly defeated.

FRUITS OF THE BIBLE

The apple has for one reason or another been regarded as an ideal fruit since ancient times. That is why in various mythologies—notably ancient Greek, Celtic, and Norse—apples may be described as "golden" and associated with the sun and youthful fertility. It is also why the fruit on the Tree of Knowledge in the Garden of Eden is generally thought of as an apple (although never described as such in the Bible), and why the Biblical expression "the apple of [one's] eye" means what it does (even if that was borrowed from an originally Babylonian figure of speech).

In the Bible, however, apples are actually not mentioned as many times as are grapes and figs. Grapes, for example, are mentioned altogether thirty times, albeit sometimes similarly figuratively (for instance, in Jeremiah 31:29 and in Ezekiel 18:2, leading to the expression "sour grapes"). Figs, on the other hand, are mentioned nineteen times, and fig trees twenty-four times—which would seem to suggest that they were of even greater significance. Adam and Eve covered their nakedness with fig leaves, Isaiah attributes medicinal properties to fig juice (specifically in treating ulcers), and (ripe) figs and fig trees seem to have been a metaphor for beauty and health. This is entirely appropriate in light of where figs tend to grow best—the eastern Mediterranean and the Near East. It is perhaps unfortunate that in temperate zones, such as northern Europe and North America, figs have the reputation of being no more than excellent laxatives.

THE DELICIOUS MONSTER

Also known as the Swiss cheese plant, the Mexican breadfruit, the ceriman, the windowleaf, and the fruit-salad plant, *Monstera deliciosa* is a creeping vine native to Central America on which the fruit, until they are ripe, look like ears of corn covered in hexagonal green scales. It takes a full year for the fruit to ripen, during which time it is dangerously toxic to humans because of the volume of oxalic acid it contains. On ripening, however, the outer green scales fall away, exposing the fleshy rind that surrounds the harder central core, and it is safe to eat. The taste is said to be somewhere between mango and pineapple, with the occasional tang of rhubarb (presumably because of the delicious monster's oxalic-acid content).

THE MYSTIC POWER OF THE LIME

In Indian tradition, the lime has the power to combat evil. It is used in tantric rituals to cleanse those possessed by evil spirits and to ward off the evil eye.

THE PAPAYA
Carica papaya

CARICACEAE

THE PAPAYA, also called the pawpaw (but not to be confused with the other fruit known as the pawpaw, see page 77, which is a North American native from the *Asimina* genus), is the bulbous melon-like fruit of *Carica papaya*. This is a tall plant with a single stem that, despite growing up to thirty-three feet high, is technically not a tree because it has no real bark. Following genetic investigation over the last few years (it was the first fruiting plant to have its genome completely decoded), *C. papaya* is now classified as the only species in its genus. It is a native to tropical regions of the New World—it cannot tolerate frost at all—but was first cultivated in southern Mexico at a time almost in human prehistory. Today, however, the papaya is commercially cultivated in many tropical and subtropical countries around the world.

The papaya initially grows with a green outer rind. As the fruit ripens, the rind changes to white, then to yellow, and then to a golden color that may have pink tinges within it. Ripeness is indicated when the rind is the right color and is very soft under thumb pressure. Inside, the flesh is pinkish-orange and surrounds a mass of black seeds. The flesh is usually eaten raw, without the seeds—although the seeds are edible, have a spicy taste, and in some parts of the world are ground up as a substitute for black pepper.

"Papaya," hand-colored engraving by Georg Dionysius Ehret (1708–1770)
from Plantae Selectae
LONDON, 1750–1773

TOMMY ATKINS, THE MANGO

The mango cultivar that is found most commonly in grocery stores and markets all over the world is called 'Tommy Atkins.' In fact, in the United States and in most countries of northern Europe, eighty percent or more of all mangoes eaten are 'Tommy Atkins' mangoes. The name is oddly appropriate in several ways. For Tommy Atkins has, since at least 1815 (when the Duke of Wellington was in charge of British military forces), been a generic name applied to the ordinary British private soldier—with particular emphasis on the word "ordinary."

The 'Tommy Atkins' mango was first produced by commercial researchers in Florida in 1940—and was initially rejected, precisely because it was so ordinary: it was not remarkably tasty or sweet, it was comparatively fibrous in texture, and there were other mango cultivars immediately available that were much better in these respects. However, like Tommy Atkins the soldier, 'Tommy Atkins' the mango turned out to have some redoubtable properties: it was prolific, it was hardy, it displayed a reasonably good appearance, and its shelflife was longer than most. And so it became popular. The fact remains that in India and Pakistan—where many kinds of mangoes are grown and enjoyed—the local variety closest to the 'Tommy Atkins' (the 'Fajri,' which is the last to fruit in the season) is used only to feed animals and is widely regarded as unfit for human consumption, although it is used as a tenderizing agent in marinades.

THE ORIGIN OF THE COCONUT

According to the legends told in Tahiti, the moon goddess Hina fell in love with an eel. Hina's brother Maui—the wise old man and bringer of fire—then killed the eel and instructed Hina to bury its head in the ground. In her distress, Hina just left the head where it was, which was beside a stream. Later, when she remembered, she came back to look for the head. It wasn't there—but in its place there now grew a magnificently handsome (and the first-ever) coconut tree.

PINEAPPLE POLLINATING

The hummingbird is the natural pollinator of pineapple plants. However, the production of the seeds that occurs after pollination negatively affects the quality of the fruit. The importation of hummingbirds into Hawaii, where (seedless) pineapples are cultivated on a huge scale and are a major export crop, is, therefore, officially prohibited. Elsewhere in the world, meanwhile, some wild pineapples rely for pollination not on hummingbirds but on species of bats, and the plants open their flowers during the nighttime accordingly and close them again at dawn.

THE CRANBERRY

Cranberries grow as bush shrubs or trailing vines in acidic marshlands in the colder regions of the Northern Hemisphere. The fruits are rounded berries that develop in clusters amid tiny leaves, and are initially white, but ripen to a deep red. Even when ripe, however, cranberries are highly acidic, tasting both sour and bitter. The most popular way to eat them is in a sauce or sweetened conserve as an accompaniment to meat dishes.

THE HOLISTIC APPLE

The flesh of the apple is held to have both astringent and laxative properties and to clean the teeth and gums while being eaten. The bark of the tree (because of the whole tree's high water content) is said to have "refrigerant and soporific" qualities; an infusion of the bark may be used to treat fevers. The leaves also have anti- bacterial properties. Apple juice reduces acidity in the stomach. The very high amount of pectin in apples, which partly explains the use of apples in combination with other fruits when making preserves (pectin helps the set), is said also to have some effect in protecting the human body against radiation.

THE NORTH AMERICAN PAWPAW

The North American pawpaw—known alternatively as the papaw, prairie banana, Kentucky banana, West Virginia banana, Ozark banana, Kansas banana, and similar—is not the same fruit as the papaya, which is also called the pawpaw. As is evident in some of its names, it is visibly a little more like a fat, squat banana than a bulbous melon and, as one of the *Asimina* genus, is related instead to the sugar apple and the custard apple of the family Annonaceae. It is the only one of that family not found only in tropical regions. The fruit ripens from green to yellow or brown on the tree, and the pulp it contains has a taste described as between banana and mango, and encloses a mass of seeds.

The pawpaw is used for baked desserts and as juice for fruit drinks; it can be used to substitute for bananas in most banana recipes on a weight-for-weight basis. However, fresh fruit may turn from unripe to overripe and already fermenting within a day, so most pawpaws are canned or frozen, or turned into preserves and jellies. The Ohio Pawpaw Festival takes place at Lake Snowden, near Albany, every September.

THE APPLE

Malus spp.

❖

R O S A C E A E

THE TECHNICAL NAME for the ordinary orchard apple is *Malus domestica*, which is closely related to *M. sylvestris*, the European crab apple, but which, following genetic investigation (the full genome was decoded in 2010), would seem to have originated as a hybrid between *M. sylvestris* and an apple species from western Asia, *M. sieversii*. However, until the 1960s the apple was in scientific terms lumped together with the pear (and similar plants to which it is indeed related) and was, thus, classified as *Pyrus malus*.

Apples as a fruit contain useful amounts of vitamin C and phenols (which help to reduce cholesterol), folic acid, and potassium (with traces of iron, magnesium and zinc, and vitamin B), and are often advertised as being particularly healthy things to eat, but it has been estimated that, in general, on a scale of one to five for medicinally beneficial properties, the apple scores a somewhat feeble two.

One thing to be noted about all members of the apple and pear genus is that the seeds (and, to a lesser extent, the leaves) contain hydrogen cyanide, so they should never be swallowed, and certainly not in any quantity. Yet it has to be said that, in minute doses, hydrogen cyanide can stimulate the breathing and improve digestion, and may even have some therapeutic properties in the treatment of some forms of cancer.

"Apple," by S. Berghuis (n.d.)
Private Collection
COLOR LITHOGRAPH BY G. SEVEREYNS (1860)

APRICOTS RAW & DRIED

Apricots are sold both as raw fruit, for eating fresh, and in dried form, either as sweet snacks or for mildly medicinal purposes (like dried plums—or prunes—they are acknowledged as effective laxatives). However, the two forms of the fruit are surprisingly different in a nutritional context. The figures in the table below are from the U.S. Department of Agriculture (2010).

NUTRITIONAL VALUE PER 100 GRAMS (3½ OUNCES)

	RAW APRICOT	DRIED APRICOT
Energy	48 cal	241 cal
Carbohydrates (Sugars)	11 grams (9 grams)	63 grams (53 grams)
Dietary fiber	2 grams	7 grams
Fat	0.4 grams	0.5 grams
Protein	1.4 grams	3.4 grams
Vitamin A equivalent	96 micrograms	180 micrograms
Beta-carotene	1,094 micrograms	2,163 micrograms
Vitamin C	10 milligrams	1 milligram
Iron	0.4 milligrams	2.7 milligrams

On this list, only vitamin C is reduced through the drying process: every other nutritional aspect is enhanced and concentrated in the dried form.

THE ANALYZED STRAWBERRY

By weight a strawberry is slightly more than ninety percent water. Its total dietary fiber is just less than two percent, and whereas the vitamin C content by weight is relatively high (more by weight, for example, than in an orange), its content of the mineral potassium by weight is actually almost three times higher (and twelve times higher than the content of its next most quantifiable mineral, calcium).

THE UNEXPECTED AVOCADO

The word "avocado"—variants of which exist in most major languages—comes from the Nahuatl *ahuacatl,* meaning "testicle," referring to its shape.

HOW TO GROW AN HEIRLOOM APPLE TREE

Follow these instructions to grow an apple tree from pits in traditional home-growing fashion. The eventual fruit is unlikely to taste (or even look) like the fruit from which the seeds come, because of the horticultural expertise that goes into producing apples for the table. However, you may be lucky.

1. Separate the seeds and put them on a flat surface for a couple of days until dry; then roll them up tightly in a moistened paper towel and refrigerate for three to four weeks, periodically checking that it does not dry out during that time (moisten if necessary). When the seeds sprout, put them in a small, shallow flowerpot filled with potting mix, and water daily; the sprouted seeds should soon become seedlings.

2. Plant each of the seedlings in a full-size pot containing potting mix and an organic fertilizer; water daily.

3. The following spring, when there is no longer any threat of frost, if your seedlings are at least three inches tall, plant them where they will get sun and have enough room to grow twenty feet tall and fourteen feet in diameter. For each one, dig a hole larger than the current root spread; gauge the depth so that the soil level of each plant in its pot is just below ground surface level. Fill in with soil (but no fertilizer), firming in with your knuckles as you work; water in well, adding more soil if the soil level sinks.

4. Spread a ring of straw, hay, or wood chips around the plant at a radius of about three feet, for protection and for moisture retention.

Care: After six to eight weeks use slow-release fruit-tree fertilizer. For the first year, water at least once a week, more often in hotter, drier periods; after that, water only during hot, dry periods.

Fruit: Wait for five years to get edible fruit and then protect against insects and diseases as required.

Pruning: Prune after the tree has first fruited, then annually in late winter or early spring to increase air circulation and sun penetration and to improve the next crop of fruit.

THE GOOSEBERRY

Ribes uva-crispa

GROSSULARIACEAE

THERE IS SOMETHING STRANGE about the gooseberry. For one thing, it is closely related to the currants—the red currant and the black currant—and yet it looks and tastes very different. The relationship is close enough to be able to produce hybrid plants. Indeed, one hybrid—the jostaberry, which is precisely a cross between a gooseberry and a black currant, and which when ripe looks and tastes like an outsize black currant—has been well established in the gardens of knowledgeable fruit growers in northern Europe (especially Germany, where it was first cultivated) for nearly 100 years. However, the European gooseberry (*Ribes uva-crispa*) is not quite the same as the North American gooseberry (*R. hirtellum*), also called the hairystem or swamp gooseberry. It is truly amazing how many other plants, mostly not only unrelated but utterly different in appearance, have been described as "gooseberries."

The original English name of the kiwifruit was the "Chinese gooseberry." The Cape gooseberry (*Physalis peruviana*), meanwhile, is also called the Peruvian cherry, but is, in fact, a relative of the tomato and the potato (family Solanaceae). The Otaheite or Malay gooseberry (*Phyllanthus acidus*) hangs in green bunches from its woody shrub; it is also edible, but strongly acidic and otherwise bears only a passing resemblance to a gooseberry. The Barbados gooseberry (*Pereskia aculeata*) is a cactus on which the berry-like fruit may be any color between white and pinkish-red.

"Gooseberry," color lithograph by E. Walther (n.d.)
from Die Pflanzen-welt
J. F. SCHREIBER, ESSLINGEN (1891)

THE SUGAR APPLE
Annona squamosa

ANNONACEAE

THE SUGAR APPLE or sweetsop is the fruit of the tree *Annona squamosa*, which is native to tropical regions of the Americas and to India and Pakistan. It is known in India and Australia as the custard apple, which is potentially confusing, because although the fruit's flesh is often described as tasting of custard, the custard apple is (outside the United States and Britain) formally the name of the fruit of the related tree *A. reticulata*, and in the United States and Britain is the name of the fruit of the equally related tree *A. cherimoya*. The sugar apple looks like a rounded, light green, somewhat lumpy pinecone. Appearances can be deceptive, because, in fact, it has a thin scaly skin surrounding a mass of glutinous and sweet white or whitish-yellow flesh that conceals a small number of scattered hard, black glossy seeds.

The sugar apple is grown commercially in many tropical and warmer subtropical countries of the world, including Thailand (where a nickname for it is "the hand grenade" due to its appearance) and the United States (in southern Florida), but in some places it has become an invasive plant and in others pollination by bees is a problem because of the tightly closed female flowers.

In the Philippines, where the fruit is eaten by a species of fruit bat, which transfers the seeds from island to island, the flesh of the sugar apple is the basis for a type of wine, which presumably does not taste of custard.

"Sugar Apple," watercolor by Charles Plumier (1646–1704)

from Plants of Martinique and Guadeloupe

NATIONAL LIBRARY OF PARIS

THE FRUCTIFEROUS ERICACEAE

The Ericaceae are known best as the heath or heather family, and are found mainly in temperate zones. It is in these zones—although there are in addition many tropical species—that the family's best-known fruit-bearing examples occur. What is most extraordinary about the Ericaceae is that many species have an essential symbiotic relationship with individual fungal species, which provide the plants with nutrients. Nonetheless, in the final total of 146 species listed, very few are in fact fruit-bearing that is in a way beneficial to humans.

Those that are beneficial include (and the list below is divided into genetically different sections even within the same genus):

Gaylussacia baccata . . . . . black huckleberry	*V. caesariense* . . . . . . .New Jersey blueberry
G. brachycera . . . . .box huckleberry	*V. darrowii* . . . . .evergreen blueberry
G. dumosa. . . . . .dwarf huckleberry	*V. myrtilloides* . . . . . . .Canadian blueberry
G. fromosa. . . . . blue huckleberry	
G. mosieri . . . . woolly huckleberry	*V. cespitosum*. . . . . dwarf bilberry
G. ursina . . . . bear huckleberry	*V. deliciosum* . . . . . . cascade bilberry/piper
	V. dentatum . . . ohelo (Hawaiian bilberry)
Vaccinium macrocarpon . . . . .American	*V. myrtillus* . . . . .blueberry/bilberry/
cranberry	whortleberry
V. microcarpon . . . . small cranberry	*V. ovalifolium* . . . . Alaska blueberry
V. oxycoccus. . . . . common cranberry	*V. parvifolium* . . . . red huckleberry
	V. scoparium . . . . grouse whortleberry
V. arboreum . . . . farkleberry/sparkleberry	*V. uliginosum* . . . .bog bilberry/blueberry/
V. crassifoleum. . . . . . . .creeping blueberry	whortleberry
V. angustifolium . . . . . .lowbush blueberry	*V. vitis-idaea* . . . . red whortleberry

THE IRRESISTIBLE BANANA

The characteristic taste of a banana is in some part due to the presence of isoamyl acetate, which is also used in flavoring pear-drop sweets—but which in addition is a seductive pheromone for honey bees.

HOW TO MAKE BLUEBERRY JAM

"Fresh" blueberries are available in supermarkets worldwide for much of the year, although frozen blueberries can also be used to make jam, once they are properly defrosted. This recipe makes close to two cups of jam or two to three jars.

Ingredients:
- 3 cups blueberries (at room temperature)
- ¼ cup lemon juice
- ¼ cup water
- 2 cups granulated sugar

1. First, sterilize two to three jam jars, depending on their size. This can be easily achieved by washing them in hot soapy water, rinsing, then placing them upside down in a cool oven at 275°F for about thirty minutes.
2. Put the blueberries, lemon juice, and water into a small to medium saucepan, bring to a boil, and let simmer for around ten minutes, stirring from time to time, until the berries have burst and the whole mixture softens.
3. Add the sugar, stirring, until it is dissolved.
4. Bring back to a boil, and let it boil for twenty to thirty minutes or until the setting point is reached.
5. Remove from the heat and scoop off any scum that rises to the top. Let stand for an additional fifteen minutes.
6. Ladle the jam into the jars and let stand to cool down.
7. When cool, seal the jars (and add labels).

THE BLUEBERRY

Vaccinium macrocarpon

E R I C A C E A E

THE BLUEBERRY IS perhaps the classic fruit of North America, although like the rest of the *Vaccinium* species (such as the cranberry), there are European and Asian species too. Altogether there are sixteen listed North American *Vaccinium* blueberries, including the American blueberry (*V. cyanococcus*), the Canadian blueberry (*V. myrtilloides*, also called the sourtop or velvetleaf), the lowbush or "wild" blueberry (*V. angustifolium*), and the northern highbush blueberry (*V. corymbosum*), which is the species most under cultivation.

Other *Vaccinium* species may also have blue berries (perhaps the best known being the European bilberry, *V. myrtyllus*). It should additionally be noted that one or two North American *Vaccinium* species are commonly locally described not as blueberries but as huckleberries. All of these, sold as "fresh," including those after transportation from North America to Europe, may in Europe be called blueberries, too. The way to tell them apart is by the color of the flesh. Blueberries (especially *V. cyanococcus*) have white or light green flesh; bilberries and huckleberries have red or purple. Blueberries have a lot of tiny seeds, whereas huckleberries have fewer, larger seeds. The wild blueberry is the "official fruit" of Maine and the "official berry" of the Canadian province of Nova Scotia; commercial rivalry between Maine and Quebec in producing lowbush blueberries has led to occasional cross-border marketing tensions.

"Blueberry," color lithograph by E. Walther (n.d.)
from Die Pflanzen-welt
J.F. SCHREIBER, ESSLINGEN (1891)

THE MANGO
Mangifera indica

ANACARDIACEAE

IF EVER THERE WAS a tree that might represent a connection between the heavens, the earth and the underworld, and that had special connotations of longevity, the common or Indian mango tree (*Mangifera indica*) could be it. It grows up to 130 feet tall; its taproot descends downward to perhaps twenty feet, accompanied by solid anchor and feeder roots; and it has been known to bear fruit at the age of 300 years. Time is evidently not of the essence to the mango, which is no doubt why its fruit may take up to six months to ripen.

The mango fruit is surprisingly variable in both size and coloration. Its skin color may be anywhere between red to green to orange and yellow. Ripe fruit exude a resinous, sweet odor. However, picking them may be a problem for some people. The mango skin, and leaves and stems, contain urushiol, the potentially allergenic substance in poison ivy that produces painful and occasionally serious contact dermatitis in humans.

Nonetheless, the Food and Agriculture Organization of the United Nations estimates annual world production of mangoes at around thirty-six million tons, so accounting for about half of all tropical fruits produced worldwide. India is the largest producer, but exports virtually none of its mango harvest, and, thus, contributes less than one percent to the international mango market trade.

"Mango," illustration by Margaret Bushby Lascelles Cockburn (1829–1928)

from Neilgherry Birds and Miscellaneous

MARGARET BUSHBY LASCELLES COCKBURN (1858)

THE ELECTRIC LEMON

Because of its high acid content, a large and juicy lemon can be made to work as a battery to produce electricity. It is an "experiment" often performed for the amazement and amusement of schoolchildren in science classes, and perfectly demonstrates that electricity can be produced from a simple battery consisting of no more than two different metals suspended in an acid solution.

Lemon juice is certainly an acid solution, generally between five percent and six percent citric acid, and two suitable different metals are easily obtainable in the form of galvanized nails (which are coated in zinc) and copper (at least partly) coins. If a zinc-coated nail is forced into one end of the lemon and a copper coin is pushed into a slot made through the zest at the other end, a cable that connects both metals to a voltmeter should show that a current is being produced between them, flowing from the zinc-coated nail (the negative electrode) to the copper coin (the positive electrode). With a single lemon, the current produced will not be great—normally less than one volt—but more lemons, connected in series, may increase the flow adequately enough to power any device that will work on a low voltage (such as a battery-operated wall clock or a digital watch). Be careful not to produce too much power for the device chosen.

KIWIFRUIT & CARDIOVASCULAR BENEFITS

Recent clinical studies in Oslo, Norway, have strongly suggested that kiwifruit contains substances that purify and thin the blood in humans, and that could, therefore, be beneficial to those who have cardiovascular problems. In a suggestion that parallels popular (but recently derided) mainstream aspirin therapy, it is said that eating two or three kiwifruit per day for four weeks has a markedly salubrious effect in lowering lipid (fat) levels in the blood, and so reduces the risk of atherosclerosis, thrombosis, heart attacks, and strokes.

THE NEW ZEALAND BUSH LAWYER

In New Zealand there is a group of five *Rubus* species of trailing blackberry vines that grow within the dense forests, and from which Maori children have for centuries delighted in eating the smallish purple fruit raw. The common English name for these blackberry plants is "bush lawyers." They are called that because the vines are covered in unpleasantly long recurved thorns that, when they get hold of you, don't let go until they have drawn blood.

BLACKBERRY®

The electronic smartphone device called the BlackBerry® was so called because at the time a name was being sought for it, the device was black and the little buttons on it resembled the drupelets of the fruit—and because, for the purposes of registering a trademark, it was essential to find a name that was memorable, verbally unconnected with terms for electronic communications (because such terms are regarded by the registering authorities as "in common usage"), and customer-friendly in other connotations. And, of course, there was the prior example of Apple.

DRUPES

Drupes are the kinds of fruit that have a large central pit that is surrounded by soft edible tissue ("flesh") all contained within a "skin" (see page 10). The technical terms for these elements are, respectively, the endocarp, the mesocarp, and the exocarp. The hard endocarp contains the seed of the plant. (A fruit in which the endocarp is soft and the seed lies exposed within it, such as an avocado, is technically a berry.) The best-known drupes are:

- almond
- apricot
- cherry
- coconut
- litchi
- olive
- peach
- pine (or fire) cherry
- plum

Also associated with drupes are those fruits that comprise aggregated "drupelets," such as blackberry, raspberry, and cloudberry, and that are, therefore, classified as polydrupes.

APPLES & PEARS

Some pears look like apples, and some apples look like pears. This is not entirely surprising because they are genetically closely related. However, there is at least one way of telling them apart other than by taste. The "flesh" of a pear contains "grit": woody concretions otherwise known as stone cells, distributed throughout the pear, but particularly around the core. Another, perhaps more impressive, difference is less immediately obvious. Placed carefully in water, an apple floats; a pear sinks. This is an extremely useful distinction if ever you are bobbing for apples, and the apple you are bobbing for turns out to be a pear.

THE LEMON
Citrus × limon

R U T A C E A E

Cultivation of the lemon—initially as an ornamental plant at least as much as for its fruit—did not really begin until the tree became popular as a feature in the Islamic gardens of Arab territories around 900 CE. It took another 550 years for it to be cultivated in Europe—first in Genoa, Italy—but within another fifty years, in 1493, Christopher Columbus was taking lemon seeds with him for planting in Hispaniola (the large island that is now occupied by the Dominican Republic and Haiti). Even so, it was not until the second half of the 19th century that the culinary and dietary uses of the fruit had overtaken the decorative use of the tree. Of those culinary uses, most centered on the juice of the lemon and on its acidic properties when used, for example, to tenderize meat (which it does by partly hydrolyzing the tough fibers of collagen) or as a marinade for fish (in which the citric acid neutralizes amines in the fish).

Figures for the world production of specific fruits by nation are annually published by the Statistical Division of the Food and Agricultural Organization (FAO) of the United Nations. Remarkably, the FAO does not distinguish in its statistics between lemons and limes, and no regular or authoritative statistics for the individual fruits are available. The current leading producers of lemons and limes are India (about sixteen percent of the world total), Mexico (fourteen percent), Argentina (ten percent), and Brazil (eight percent).

LIMONIER SAUVAGE.

Limone sylvatico.

Tab.70.

Poiteau pinx.   *Gabriel sculp.*

"Wild Lemon," engraving by Gabriel Sculpt (n.d.)
from Histoire Naturelle des Orangers
WATERCOLOR BY PIERRE-ANTOINE POITEAU (1766–1854)

TYPES OF PLUMS

The botanical genus *Prunus* comprises a number of subgenera, one of which is also called *Prunus*. That subgenus is in turn categorized, according to most authorities, in three sections, one of which is yet again known as *Prunus*, whereas the others are *Prunocerasus* and *Armeniaca*—the last of which some commentators insist is, instead, another subgenus. The chart below may or may not make things easier to follow.

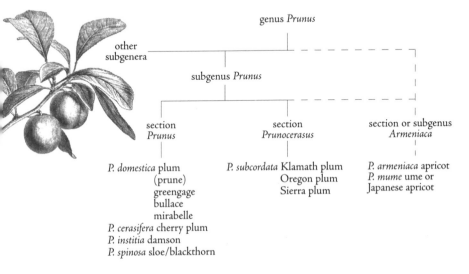

genus *Prunus*

other subgenera

subgenus *Prunus*

section *Prunus*	section *Prunocerasus*	section or subgenus *Armeniaca*
P. domestica plum (prune) greengage bullace mirabelle *P. cerasifera* cherry plum *P. institia* damson *P. spinosa* sloe/blackthorn	*P. subcordata* Klamath plum Oregon plum Sierra plum	*P. armeniaca* apricot *P. mume* ume or Japanese apricot

LEMON JUICE

It is the citric acid content of lemon juice that gives it such a sour taste (and a pH of between two and three). The average lemon contains about three large tablespoonfuls of juice. Ensuring that the lemon is at room temperature (or bringing it to room temperature in a microwave oven) renders the juice easier to extract.

The uses of the juice are not merely culinary. Lemon juice can lighten hair color, can substitute for short-term preservatives in treating raw foods that oxidize and turn brown on continued exposure to air (such as sliced apple), and, when mixed with baking soda, can remove stains from plastic containers. In south and east Asia, lemon juice has been used for centuries as an antiseptic and even as an effective antidote to certain vegetable poisons.

HOW TO MAKE LEMON CURD

Lemon curd is surprisingly easy to make. Its egg contents means that, unlike jam, it does not have to be rapidly boiled to set. In fact, the trick is to heat it gently for the perfectly smooth texture. Once opened, keep the jar in the refrigerator. This recipe makes close to three cups of jam.

Ingredients:
- 4 large lemons
- 4 large eggs
- 1¾ cups superfine sugar
- 1 cup (2 sticks) unsalted butter, at room temperature
- 2 teaspoons cornstarch

1. First sterilize three to four jam jars, depending on size. This can be easily achieved by washing them in hot soapy water, rinsing, then placing them upside down in a cool oven at 275°F for about thirty minutes.
2. Zest and juice the lemons.
3. Whisk the eggs lightly in a medium saucepan using a balloon whisk.
4. Add all the remaining ingredients, and place the saucepan over a medium heat, whisking continuously with the balloon whisk for about eight minutes until all the sugar has dissolved, the butter is incorporated, and the mixture thickens.
5. Reduce the heat to low and let simmer for one minute, still whisking continuously.
6. Remove the saucepan from the heat, and pour the mixture into the jars, filling each to the very top and covering each at once with a disc of wax paper. Seal the jars while still hot.
7. When the jars are cold, add labels.
8. Store in a cool place (and eat within four or five weeks).

THE APRICOT

Prunus armeniaca

ROSACEAE

THE APRICOT WAS thought to have originated in Armenia, which is why its scientific name is *Prunus armeniaca*. However, it now seems more likely that the fruit actually comes either from China or India. In any event, it is traditionally believed that apricots were introduced to Europe at the behest of Alexander the Great (356–323 BCE). Today, apricots are grown anywhere there is a fairly dry climate and cold winters (apricots require an annual period of dormancy). In the United States most apricot cultivation is in California; in Australia it is in South Australia. These are both areas that get few frosts in springtime— spring frost kills apricot flowers. Elsewhere, another possibility is to hybridize apricots with plums (which can be done successfully) to suit slightly different environments; the resulting fruit are known variously (in English) as "pluots," "plumcots," or "apriums."

Apricot seeds or kernels, removed from the hard shell often called the "pit," are regarded as acceptable substitutes for almonds in many of the countries that surround the Mediterranean, and from there to central Asia. The Italian liqueur amaretto is also flavored with extract of apricot kernel. However, as in many fruit seeds and pits, major constituents of these kernels are glycosides, which may cause the production of hydrogen cyanide within the body.

"Apricot," color copper engraving by Pierre Joseph Redouté (1759–1840)
from Choix des Plus Belles Fleurs
RETOUCHED BY LANGLOIS

THE STRAWBERRY
Fragaria vesca

❧

ROSACEAE

IT WAS THE ANCIENT ROMANS who first cultivated strawberries. They called them *fraga*, literally "fragrant things," from which the scientific term for the genus *Fragaria* is derived. (The Romans also called raspberries *fraga ambrosia* "ambrosial strawberries," from which modern French *framboise* is derived.) Today, most strawberries generally exude little scent. That may well be because, after the Romans, no one cultivated strawberries in a consistently serious fashion until the 17th century in northern Europe, and then the species involved was the woodland or alpine strawberry *Fragaria vesca*. Finally, in 1740 in Brittany, northwest France, what is now known as the garden strawberry was crossbred; it is a hybrid between the flavorful North American *F. virginianum* and the outsize Chilean *F. chiloensis*, which is currently cultivated worldwide. For commercial production, however, to be used in flavorings and in canned preparations, the modern standard strawberry is *F.* × *ananassa*.

In 2007 the world leaders in strawberry production were: Canada (1.6 million tons), the United States (1.2 million), Spain (501,000), Russia (324,000), and Turkey (239,000). North America was also where the most strawberries were eaten: U.S. consumption of strawberries in 2006 was about three-and-a-half pounds per resident, and about one-and-three-quarter pounds of frozen strawberries per resident.

"Late Scarlet Strawberry," color copper engraving by William Hooker (1779–1832)

from Pomona Londinensis

THE HORTICULTURAL SOCIETY OF LONDON (1818)

HUCKLEBERRIES

What is and is not a huckleberry? It is by no means always clear. In the United States and Canada, some huckleberries (or dangleberries) are what are elsewhere called blueberries—which is not entirely surprising because blueberries are *Vaccinium* spp., whereas huckleberries can be *Vaccinium* spp. (as, for example, the red huckleberry, *V. parvifolium*), but can be (and are mostly) *Gaylusaccia* spp., and the only real physical difference is that by tradition blueberries have multiple tiny seeds whereas huckleberries have ten larger (and less swallowable) seeds.

Many people claim that there is also a distinction between the flavors of blueberries and of huckleberries. Scientifically, however, there remains contention as to whether *Vaccinium* and *Gaylusaccia* should be classed as two separate genera at all. Nonetheless, there are eight recognized types of North American *Gaylusaccia* huckleberry, including the bear huckleberry (*Gaylussacia ursina*), the blue huckleberry (*G. frondosa*), the hairy-twig huckleberry (*G. tomentosa*), and the woolly huckleberry (*G. mosieri*). The remaining forty-two species that are currently identified as of *Gaylussacia* are native to mountainous areas in South America, where the *uva de Páramo* (the "Páramo grape," *G. buxifolia*) of Venezuela and Colombia is perhaps the best known. In the United States the word "huckleberry" at one time described anything that was particularly small or nice to have around. Presumably, this is why Mark Twain's *Huckleberry Finn* (1884/5) was so named.

CHERRY RED

John Gay's (1688–1732) libretto for *Acis and Galatea*, set to music by George Frideric Handel (1685–1759), includes the aria "[O] Ruddier than the Cherry," which supposedly describes the lady love of the jealous giant Polyphemus, who is about to rob the nymph Galatea of her love and lover Acis. It shows not only how the color red was understood to be a sign of both beauty and health in 18th-century England, but that it was associated primarily with cherries (and not apples, strawberries, raspberries, or tomatoes, much less sweet peppers or cooked beet).

THE SYMBOLIC QUINCE

In Slavonia, Croatia, when a baby is born it is traditional to plant a quince tree as the symbol of a new life, a new love, and an extended future for the family. For the ancient Greeks, the quince was sacred to their goddess Aphrodite, and the fruit was a ritual offering at weddings.

HOW TO GROW HEIRLOOM STRAWBERRIES

Strawberries enjoy a sunny, out-of-the-wind spot that is preferably on higher ground, where there is least risk of spring frost. Raised beds are recommended. Avoid establishing a strawberry bed where potatoes, tomatoes, sweet peppers, or grass have already been grown or are near. Buy as plants from a garden center or nursery; do not try to grow from seeds. Plant in late summer, unless frost is likely; otherwise, early spring. "Perpetual strawberries," which produce three "flushes" of fruit in a summer, should always be planted in early spring. Propagate from the second year (see below).

1. One month before planting, dig in a good quantity of organic matter (compost) plus two handfuls of bonemeal per square yard.
2. Add general fertilizer to the bed four or five days before planting.
3. Plant to the same depth as they were in the pots; it is essential not to plant too high (the root will dry out) or too low (the root will rot). Water well, but be careful never to waterlog.
4. Once established after about three weeks, stop watering until the fruits appear, then water again.
5. Varieties planted in spring should have the first blooms pinched off to consolidate the roots. There should be no need to feed the plants if a layer of compost is spread around the plants in early spring (but if you do feed, do so when fruits are forming and use high-potassium fertilizer).

Fruiting: When the fruits appear, the plant will also produce runners with nodes (representing next year's crop): remove them to increase the current year's yield (or leave a few to propagate from). When the fruits are large enough to nearly touch the ground, place a layer of straw or black plastic sheet between the fruit and the ground to prevent the fruit from rotting;—if using plastic, make multiple drainage holes through it. Protect the fruit from birds with wire mesh or plastic netting secured by stakes.

Propagating for next year: Let a runner root at a node; three weeks after it has rooted, cut the runner, dig up the plantlet and the soil around it, and place in a pot with potting soil.

THE QUINCE
Cydonia oblonga

ROSACEAE

THE QUINCE TREE (*Cydonia oblonga*) is native to the Caucasus region of southwest Asia. However, the current leader in producing quinces is Turkey, which is actually responsible for one quarter of the overall total world harvest.

It is possible that, historically, the cultivation of the quince came before the cultivation of the apple—to which the quince is related. It is also possible that many of the ancient scriptural and mythological references to apples, and especially golden apples, were originally intended as references to the quince, which when ripe is distinctly golden yellow—much more so than most apples. The one major drawback to this suggestion, however, is that whereas apples were prized throughout history as a flavorsome fruit that could be eaten raw, most quinces are too hard and too sour to eat raw until they have been "bletted" (allowed to begin the decaying process). Yet once peeled, they can be baked or stewed or even roasted, and they can be turned into jelly or jam. In fact, the word marmalade originally referred to quince jam (from Portuguese *marmelo*, "quince"), and because they contain proportionately more pectin even than apples, quinces are exceptionally useful as additional ingredients when making other fruit jams and getting them to set. Quince juice, meanwhile, is a favourite cold drink in Germany. A quince brandy is also manufactured in Germany, Serbia, and Luxembourg.

"Quince," color etching by Friedrich Guimpel (1774–1839)

from Lichen Camphor, Vol. 4

F. G. HAYNE (1816)

POMEGRANATE JUICE

The pomegranate has never been popular as a fruit outside Mediterranean countries and from the Near East to India, despite its long cultivational history and its formidable list of mythological associations. In fact, it has never been commercially grown in northern Europe—and in Latin America north to Southern California it is grown almost entirely for its juice instead of what it was in former times most prized for: all the edible seeds in their seed casings (arils). The relatively modern worldwide popularity of the juice is partly due to the spread of Persian and Indian cuisines, which feature it both as a drink and, in thick and sweetened form, as grenadine syrup, used in that context primarily as a cooking sauce (but in other contexts as a mixer for alcoholic drinks). However, its popularity is partly due also to the marketing of the juice by its producers—perhaps to the point of hyperbole, and on one occasion in 2010 to the point of an official warning by the U.S. Food and Drug Administration—as having particularly beneficial medicinal properties in relation to the care and potential cure of cardiovascular disorders and deficiencies. Tests are currently ongoing as to whether the juice has in addition antiviral and/or antibacterial effects that might be of similarly remedial medicinal use.

HOT BANANAS

When the radiation monitors used at some U.S. seaports to detect illegal nuclear shipments in transit start sounding their sirens and flashing their lights, officials may—before they frantically call in the military authorities—first take a look to check whether another cargo of bananas has just arrived.

Bananas have a high potassium content, and a small part of that content is the isotope potassium-40—which is very slightly radioactive. In a single banana, the level of radiation is so low that it is virtually undetectable (and is harmless to humans), but a large cargo of bananas can trigger the monitoring alarms.

RASPBERRY KETONE

Raspberry ketone is a compound of natural phenols that together correspond to the attar, or aromatic essence, of raspberries. Because from two pounds three ounces of raspberries it is possible on average only to extract nine one-hundred-thousandths of an ounce of ketone, it is among the most expensive ingredients used in perfumery and the food industry today—at $622 an ounce—even when prepared through chemical intermediates.

HOW TO GROW HEIRLOOM RHUBARB

Rhubarb does not like very high or low temperatures (although after the first year it requires some winter frost to produce good stems), so select a cool, open site with some sunshine and perhaps some shade that has well-drained soil, ideally of pH 6.0–6.8. Buying year-old rhubarb crowns from a garden center or nursery is recommended, but you could grow your own from seeds in pots indoors over the winter and then plant outside (and wait another year or more for edible stems). Early spring is the best time to establish the crown, when overnight temperatures of down to forty degrees Fahrenheit should arouse the plant from winter dormancy.

1. Four weeks before planting, dig in a lot of well-rotted compost to some depth, and add fertilizer; be careful that all weeds are removed from the area because rhubarb doesn't like competition.
2. Plant crowns to a depth of about four inches below the soil surface—any young shoots already on a crown should appear just above the surface—and space crowns out between two and three feet apart (depending on variety: consult your garden center). If you grow seedlings, space them three to four feet apart. Firm the soil down around either form.
3. Fertilize twice a year, in early spring and late fall; water well, but sensibly—rhubarb needs less watering than most edible plants.
4. Place a mulch or nonmanure compost over the crowns in winter and, after the first hard frost of winter, remove all remaining stems.

Harvesting: Do not harvest the stems in a crown's first year (or a seedling's second year) because the roots need the nutrients in them to be retained for the following year. The next year be careful to harvest only two or three stems per plant; thereafter, harvest at will—always doing this before mid-summer and leaving at least four stems on each plant (so enabling each crown to continue producing stems for between seven and fifteen years). To harvest, beginning in spring, either twist off complete stems or cut at ground level.

Pruning: You may want to thin out the crowns after four or five years if the bed is becoming too full or if you want to use part of a crown to propagate another rhubarb plant.

THE GRAPE
Vitis vinifera

VITACEAE

THE CLASSIC GRAPE grows in clusters (bunches) of between six and three hundred, and starts as green, but turns pink and then purple as it ripens. The so-called "white" grapes—which, in fact, merely stay the original green—are actually subject to a double genetic mutation that eliminates the production in the grapes of anthocyanins: pigments that are responsible for the purple color.

It seems to have been the ancient Egyptians who first cultivated the (purple) grape, followed by the ancient Greeks, the Phoenicians, and the Romans, growing them both for eating and for making wine.

Today, the distinction between grapes for eating (table grapes) and grapes for winemaking is much more marked. Although virtually all grapes consumed by humans belong to the species *Vitis vinifera*, modern table grapes and wine grapes are significantly different as a result of selective breeding. Table grapes are generally larger, and many of them are seedless; they also tend to have thin skins. Wine grapes are smaller, mostly contain seeds, and have fairly thick skins, which are desirable for winemaking because the skins add both taste and aroma to the wine. Wine grapes are usually also much sweeter: they are harvested when their juice is around twenty-four percent sugar by weight. The juice of table grapes, on the other hand—from which cartons of grape juice are commercially prepared—is generally around fifteen percent sugar by weight.

"Grapes," color lithograph by Alexis Kreder (1839–1912)
Private Collection
PARIS (1890)

BLACKBERRIES & HEALTH

Blackberries have long been associated with relieving ill health, particularly if the fruit is picked and consumed straight from the bush. By long tradition, they are said to be efficacious in the treatment of coughing and other respiratory ailments, mouth inflammations and diarrhea. In the form of blackberry tea, the steeped fruit and leaves contain useful proportions of vitamins C and E and the mineral selenium, and are thought to be beneficial in reducing overly high blood-sugar levels. Moreover, the U.S. Department of Agriculture and other concerned bodies have in recent years been studying the potential uses of the chemical compound cyanidin-3-glucoside, which is found naturally in blackberries, in an attempt to identify and isolate cancer-inhibiting substances.

Of added benefit, the structure of the blackberry makes it among the highest fiber-content of all fruits. Like other "aggregate" fruits, it is composed of many individual drupelets, each one a small berry, that contribute additional skin, seeds, and pectin to our diet. Digestive and nutritional values apart, wild or cultivated blackberries are a delicious treat. One particularly successful hybrid, bred from two cultivars of evergreen blackberry in Marion County in Oregon's Willamette River Valley, is the marionberry, which is highly esteemed for its shape, sweetness, and superb flavor.

WINTERGREEN

The American wintergreen (*Gaultheria procumbens*), also called Eastern teaberry, checkerberry, and boxberry, is unusual for the *Gaultheria* genus in that its fruit is edible straight from the plant, which is a low-growing shrub. From a distance the fruit looks like a smallish red cherry; however, it is not a drupe but a dry capsule surrounded by a fleshy calyx.

Eaten raw, wintergreen fruit have a minty taste. An oil ("oil of wintergreen"), prepared by steeping the fruit for a period in warm water, concentrates and preserves this taste, and is used in many commercial consumer applications that have a minty flavor, such as toothpastes, chewing gum,

mouthwashes, and "minty" confectionery. The oil additionally has medicinal qualities that are used therapeutically, either within topical salves or in aromatherapy in the treatment of muscle and joint pain and circulatory and skin disorders.

Undiluted, the oil is of a surprising potency (and should not be swallowed other than under professional medical supervision): two tablespoonfuls of oil of wintergreen is equivalent to almost two ounces of aspirin (or 171 aspirin pills). Another cautionary note: the medically active ingredient in the oil is methyl salicylate, to which some people may be allergic in some degree.

RAISINS, CURRANTS & GOLDEN RAISINS

All three of these are forms of dried grape (or "dried vine fruit," according to official European Union terminology). A raisin is the basic form: it is simply a dried grape (and, indeed, *raisin* is the French for "grape"). A currant is the dried form of the grape variety called 'Zante Black Corinth' (and *currant* is a corruption of the name "Corinth"; there is no direct connection with other fruits known as currants, such as black currants). A golden raisin was originally a similar dried form of a specific Turkish grape variety, but golden raisins have for a long time now been no more than raisins chemically treated to look and taste like the older form.

KUMQUATS

Kumquats are like little olive-shaped oranges, and according to many authorities are, indeed, very closely related to citrus fruit—although other authorities insist that they belong only to their own genus, *Fortunella* (named after Robert Fortune of the London Horticultural Society, who introduced them to Europe in 1846).

The most unusual thing about the kumquat is that although it is generally eaten fresh (that is, raw), it is either eaten whole—the peel and the internal flesh together—or the sweetish peel is eaten and the sourer, saltier flesh is thrown away. Otherwise, kumquats may be used in preserves and jellies, or candied, or used as replacements for olives in salads and as cocktail garnishes. Because of their close relationship with citrus fruit, kumquats can be crossbred with them to produce any of a number of now well-recognized hybrids, such as the limequat, the mandarinquat, the orangequat (the last is actually a hybrid with the satsuma mandarin) and the sunquat (hybrid with the lemon)—in all of which the word element *-quat* represents a form of the Chinese for "orange" (*kumquat* itself corresponds to "golden orange").

THE RASPBERRY

Rubus idaeus

ROSEACEAE

RASPBERRIES ARE GROWN commercially in all temperate regions of the world. Most of them are hybrids of cultivars between *Rubus idaeus*, the European red raspberry, and *R. strigosus*, the American red raspberry. Mutation of either of these, involving the effect of recessive genes, has been known to result in the appearance of yellow or golden raspberries. Much less well known are the blue raspberry (*R. leucodermis*), which is actually dark purple, the black raspberry (*R. occidentalis*), which is dark purple to black (but tastes nothing like a blackberry), and the Arctic raspberry (*R. arcticus*), which is at least red, if small, but renowned in western Russia as "the berry of kings" and prized as a particular delicacy from Scandinavia to northern Canada.

An even more unusual form of raspberry is *R. phoenicola*, which is native to China, Japan, and Korea, and in the West is generally called the wineberry. The fruits of this plant, as they develop, are surrounded by a protective calyx covered in hairs, from which a sticky fluid is exuded that is apparently no more than a slippery barrier against insect predation—it has no special astringent or acidic properties harmful to insects. However, it does mean that, like other raspberries, the plant can become territorially invasive (as it has, indeed, become in areas of both Europe and North America).

"Raspberry," color copper engraving by Pierre Joseph Redouté (1759–1840)

from Choix des Plus Belles Fleurs

RETOUCHED BY LANGLOIS

HONEYBERRIES OR HASKAPS

Honeyberries are not particularly well known, although as fruit they are held to be delicious and medicinally they are regarded as highly efficacious. Originally a Russian subarctic relative of honeysuckle (*Lonicera* spp.), cultivars of the honeyberry or "haskap" bush are available primarily from the University of Saskatchewan in Canada, although it is said that individual cultivars should be matched to specific zonal climates (advice on which cultivars is freely available online from the U.S. Department of Agriculture).

A warning to gardeners: the plant would seem to be highly invasive, and will grow just about anywhere, because temperature, at least, is relatively insignificant. Honeyberries do particularly well where the soil is loose and well drained with a reasonable organic content: ideal pH is 6.5, but these plants can tolerate from 5 to 8, depending on the cultivar. Honeyberries fruit early in the season—before strawberries, for example—but they do so from the very first season, perhaps for another twenty-five seasons or more…

FRUITY AS A NUTCAKE

Two cultures separated by a common language—and by their idea of what goes into a fruitcake. In the United States the fruit in a fruitcake (one word) may include golden raisins, lemon or orange zest, and candied fruit pieces, in addition almost always to slivered or chopped nuts. In Britain the fruit in a fruit cake (spelled as two words) is mostly currants with a few crystallized cherry halves. The inclusion of nuts is the basis for the American description of some wilfully eccentric person as "nutty as a fruitcake"—which seems bizarrely inappropriate to many British English-speakers.

OJ IN THE DEEP FREEZE

In July 2010 there were almost one-and-a-quarter billion pounds of frozen orange juice in cold storage in the United States, which sounds like a huge quantity, although that total was down by twelve percent on the previous year.

HOW TO GROW HEIRLOOM RASPBERRIES

Raspberries are tolerant of some shade, but prefer a site where there is on average a few hours' sunshine a day. The soil should be light to medium and slightly acidic; moisture is essential, but continuous rain or overwatering endangers the plants. Buy plants as one-year-old canes (primocanes), usually in bundles with their roots bare, from a garden center or nursery, and make sure you know the fruiting season—early, mid-, or late summer, or early, mid-, or late fall—of the particular raspberry cultivars you are buying. Late fall is the best time to plant, but you can plant any time up to early spring unless the soil is frozen or waterlogged. Do not plant where potatoes, tomatoes, or peppers have already been grown or are near.

1. At least three weeks before planting, dig in a considerable quantity of well-rotted manure and/or compost, while also clearing all weeds from the site.
2. Planting depth is critical: try to use the original soil line (if visible) on each cane as a guide to the advised ground level. Usually this means a hole to be dug per cane about ten inches wide and three inches deep. Spread out the roots; fill in the hole with soil, firming with the knuckles, to a height preferably just above the original soil line.
3. Support the raspberries. In a small plot with just a few canes, support two canes or more with a single sturdy stake; in larger plots (and depending on the cultivar), a system of posts with wires strung between them at various heights above the ground may be necessary, and canes should be planted at least sixteen inches apart. Use garden string to attach plants to supports at various heights as they grow. Water well and often, but do not overwater.
4. Apply a general-purpose granular fertilizer on top of the soil every spring, and mulch with well-rotted manure or compost.

Fruiting: From the second year, usually the year in which fruit first form, harvest the fruit individually when ripe and firm; eat within a day of picking, unless preserving them.

Pruning: At least a month after the end of the fruiting season, prune canes that have fruited right down to ground level. Top and thin out others; also check for signs that root growth is not extending underground beyond your preferred limits.

THE PEAR
Pyrus communis

ROSACEAE

THE FRUIT ORDINARILY known as the pear is technically not the fruit at all but a greatly distended "calyx tube" (the top end of the flower stalk) that surrounds the real fruit—which is the core and is generally discarded. The flesh of the pear is, therefore, cellular and fibrous, instead of soft and juicy, resembling the flesh of the apple, to which the pear is closely related. There are about thirty main species of pear, all of which have subspecies, plus innumerable hybrids between species and subspecies. However, commercially grown pears belong almost entirely to three species only: the European pear (*Pyrus communis communis*, grown mostly in Europe and North America), the Asian or Nashi pear (*P. pyrifolia*, grown mostly in the Far East), and the Chinese white pear (*P.* × *bretschneideri*, grown, indeed, in China).

The pear tree's preferred habitat is a cool temperate climate—which is also a preferred habitat among humans—so the pear has been cultivated since very ancient times across Asia, Europe, and North Africa. Pears are mentioned in Homer's works, were eaten cooked (and never raw) by the Romans, and were brought by the French to the court of King Henry III of England in the mid-13th century. At that time the French may or may not have brought with them the custom also of smoking pear leaves—a practice that continued until the introduction of tobacco there three centuries later.

"Pear," illustration by Godard (n.d.)
Private Collection
COLOR LITHOGRAPH BY G. SEVEREYNS

Table of Latin & Common Names

The following entries act as a quick reference guide to the scientific names (family, genus, species) and some of the common names for the most popular fruit.

FAMILY **Anacardiaceae**

Genus	Species	Common name(s)
MANGIFERA	caesia	Malaysian mango, binjai
	indica	Common/Indian mango
	sylvatica	Himalayan mango
SPONDIAS	dulcis	Tahitian apple, pomecythere
	tuberosa	Brazil plum, imbu
ANACARDIUM	occidentale	Cashew (cashew apple)
SCLEROCARYA	birrea	Marula, jelly plum
PISTACIA	vera	Pistachio (nut)
BUCHANANIA	lanzan	Charoli/chirauli (nut)
	obovata	(Australian) green plum
CHOEROSPONDIAS	axillaris	Lapsi, candy tree

FAMILY **Annonaceae**

ANNONA	squamosa	Sugar apple, sweetsop
ASIMINA	triloba	(North American) pawpaw

FAMILY **Arecaceae**

COCOS	nucifera	Coconut

FAMILY **Bromeliaceae**

ANANAS	comosus	Pineapple

FAMILY **Caricaceae**

CARICA	papaya	Papaya, papaw, pawpaw

FAMILY **Cucurbitaceae**

CITRULLUS	lanatus	Watermelon
CUCUMIS	melo	Muskmelon, cantaloupe
	metuliferus	Horned melon

FAMILY **Ericaceae**

ARBUTUS	unedo	Arbutus, strawberry tree
ARCTOSTAPHYLOS	alpina	Alpine bearberry

Genus	*Species*	*Common name(s)*
ARCTOSTAPHYLOS	*manzanita*	Common manzanita
	rubra	Red bearberry
	uva-ursi	(Common) bearberry
EMPETRUM	*earnesii*	Rockberry
	nigrum	Crowberry
GAULTHERIA	*procumbens*	Wintergreen
GAYLUSSACIA	*baccata*	Black huckleberry
	brachycera	Box huckleberry
	buxifolia	Uva de Páramo
	dumosa	Dwarf huckleberry
	frondosa	Blue huckleberry
	ursina	Bear huckleberry
LEDUM	*groenlandicum*	Labrador tea
VACCINIUM	*macrocarpon*	American cranberry
	microcarpum	Small cranberry
	oxycoccus	Common cranberry
	arboreum	Sparkleberry, farkleberry
	angustifolium	Lowbush blueberry
	corymbosum	Highbush blueberry
	crassifolium	Creeping blueberry
	myrtilloides	Canadian blueberry
	cespitosum	Dwarf bilberry
	myrtillus	Bilberry, blaeberry
	uliginosum	Bog/northern bilberry
	ovatum	California huckleberry
	parvifolium	Red huckleberry
	stamineum	Deerberry
	vitis-idaea	Lingonberry, whortleberry

FAMILY Grossulariaceae

RIBES	*aureum*	Buffalo/golden currant
	divaricatum	Spreading gooseberry
	lacustre	Swamp gooseberry
	laxiflorum	Trailing black currant
	nigrum	(European) black currant
	quercetorum	Rock/oak gooseberry
	rubrum	Red currant
	triste	Swamp redcurrant
	uva-crispa	(European) gooseberry
	velutinum	Desert gooseberry
	viscosissimum	Sticky currant

FAMILY Malvaceae

DURIO	*dulcis*	Red durian, lahong
	grandiflorus	Durian munjit
THEOBROMA	*cacao*	Cocoa tree, cacao

FAMILY Moraceae

Genus	Species	Common name(s)
ARTOCARPUS	altilis	Breadfruit, camansi
	heterophyllus	Jackfruit, nangka
	treculianus	African breadfruit
BROSIMUM	alicastrum	Breadnut
BROUSSONETIA	papyrifera	Paper mulberry, kalivon
FICUS	benghalensis	(East) Indian or Bengal fig
	carica	Common fig
	citrifolia	Wild banyan, shortleaf fig
	religiosa	Sacred fig, bo-tree fig
	sycomorus	Sycamore fig, fig-mulberry
	virens	White fig, pilkhan
MORUS	alba	White mulberry
	nigra	Black mulberry
	rubra	Red mulberry

FAMILY Musaceae

MUSA	acuminata, balbisiana	Banana, apple banana
	x paradisiaca	Plantain

FAMILY Oleaceae

OLEA	europaea	Olive

FAMILY Polygonaceae

RHEUM	rhabarbarum	Rhubarb

FAMILY Rosaceae

MALUS	domestica	Apple, orchard apple
	sylvestris	Crab apple, wild apple
PYRUS	communis	Pear
	communis pyraster	Wild European pear
CYDONIA	oblonga	Quince
ERIOBOTRYA	japonica	Loquat, Japanese medlar
ROSA	(various)	Rose hips
MESPILUS	germanica	Common medlar
PRUNUS	cerasifera	Cherry plum
	domestica	Plum, prune
	domestica italica	Greengage
	institia	Damson
	spinosa	Sloe, buckthorn
	armeniaca	Apricot
	avium	Wild cherry, sweet cherry
	cerasus	Sour cherry
	fruticosa	European (dwarf) cherry
	padus	Bird cherry
	pensylvanica	Fire cherry, pin cherry

FAMILY **Rosaceae** *continued*

Genus	Species	Common name(s)
	tomentosa	Chinese dwarf cherry
	virginiana	Chokecherry
	caroliniana	Carolina cherry laurel
	laurocerasus	English or cherry laurel
	persica	Peach, nectarine
	dulcis	Almond
RUBUS	*caesius*	European dewberry
	chamaemorus	cloudberry
	fruticosus	blackberry
	glaucifolius	San Diego raspberry
	hispidus	Swamp dewberry
	idaeus	European red raspberry
	× *loganobaccus*	Loganberry
	strigosus	American red raspberry

FAMILY **Rutaceae**

CITRUS	*aurantifolia*	Key lime, Omani lime
	× *aurantium*	Bitter or Seville orange
	× *bergamia*	Bergamot orange
	× *clementina*	Clementine
	hystrix	Kaffir lime
	× *latifolia*	Persian lime, Tahiti lime
	× *limon*	Lemon
	maxima	Pomelo, shaddock
	× *paradisi*	Grapefruit
	reticulata	Mandarin orange
	× *sinensis*	Sweet orange
	× *tangelo*	Tangelo
	× *tangerina*	Tangerine
	× *unshiu*	Satsuma, mikan
(FORTUNELLA)	*japonica*	Kumquat
AEGLE	*marmelos*	Bael, Bengal quince
LIMONIA	*acidissima*	Curd fruit, wood apple
TRIPHASIA	*trifolia*	Limeberry

FAMILY **Sapindaceae**

LITCHI	*chinensis*	Litchi, lychee
NEPHELIUM	*lappaceum*	Rambutan
DIMOCARPUS	*longan*	Longan

FAMILY **Solanaceae**

LYCIUM	*barbarum/chinense*	Wolfberry, goji berry

FAMILY **Vitaceae**

VITIS	*vinifera*	Wine or European grape

Glossary

⚘ Anthocyanins

Water-soluble, odorless, and almost tasteless pigments occurring in the tissues of plants, coloring them red in acidic environments, blue in alkaline. As colorants, they contribute to attracting potential plant pollinators, and can protect external plant tissues against harmful sunlight.

⚘ Antioxidants

Molecules that inhibit the oxidation of other molecules by offering themselves up for oxidation, so potentially protecting cells in the body from damage by disease or aging. Vitamins C and E are antioxidants, as are flavonoids (*q.v.*).

⚘ Cross-fertilization

Also known as cross-pollination, the transfer of pollen from one plant to another, either by insects or human intervention.

⚘ Cultivar

A cultivated variety of a plant that has distinct characteristics.

⚘ Drupe

Type of fruit classically consisting of an outer skin (exocarp) containing a fleshy interior layer (mesocarp) surrounding a husk or shell (endocarp) that itself encloses a seed; the endocarp and seed together (which form the pit or stone) are derived from the ovary wall of the flower.

⚘ Drupelet

One of the small individual drupes that combine together in a characteristic fashion to form an aggregate fruit such as a raspberry or a blackberry (in which the drupelets form from a single flower) or a mulberry (in which the drupelets form from separate flowers).

⚘ Endocarp

In a fruit, the layer that encloses the seed or seeds; it is often hard and shell-like (as in peaches and plums) but may alternatively be softer and fibrous (as an apple core) or membranous and segmented (as the "flesh" of citrus fruits).

Exocarp

The outer surface, skin, rind, or peel of a fruit, which usually contains pigments (such as anthocyanins (*q.v.*)) and essential oils; in citrus fruits the exocarp (or "flavedo") further contains steroids, fatty acids, waxes, enzymes, and the terpene hydrocarbon limonene.

False berry, false fruit

Berry or fruit in which the flesh that is eaten is derived not from the plant's ovary but from some other nearby tissue; the flesh of pomes (such as pears), for example, surrounds the true fruit, which is the core; pineapple flesh is derived from tightly compacted flower berries.

Flavonoids

Plant pigments, such as anthocyanins (*q.v.*), that when consumed by humans may (or, according to some reputable scientific authorities, may not) have highly beneficial medicinal effects because of their antioxidant (*q.v.*) content; body metabolism of flavonoids is, in any case, poor.

Genus

Second most detailed classification (after species (*q.v.*)) of a plant, determining the botanical group within the "family" to which it belongs. The plural form is genera.

Heirloom

Term most commonly used in the United States to describe surviving old, open-pollinated (*q.v.*) cultivars of plants. Most pre-date 1950 and are not part of large-scale commercial agricultural production.

Hesperidium

Type of fruit of which citrus fruits are characteristic: a thickish oily peel encloses structured segments (carpels) of fluid-filled tubules that correspond to the inner ovary wall, within a layer of spongy pith that corresponds to the outer ovary wall.

ꙮ Hypanthium

Structure in a plant that results from the fusing together of the bases of the sepals, petals, and stamens; the flesh of an apple develops from the hypanthium, whereas the true fruit is the core; the fruits of cacti are mostly formed from the hypanthium.

ꙮ Mesocarp

In a fruit, the central layer between the outer exocarp and the inner endocarp, in many fruits (such as apples and pears) constituting the flesh that is eaten; in a coconut, however, it is the hard shell between the hairy outside and the inner surface to which the white coconut "meat" adheres.

ꙮ Multiple fruit

Individual fruit that develops from a cluster of flowers, each of which produces an initial form of fruit that combines with the others to finally constitute a single, larger characteristic fruit; examples are pineapples, mulberries, breadfruit, and figs.

ꙮ Nondialyzable

Describing a compound substance (especially a liquid) in which some or all of the constituent elements cannot for one reason or another be separated out or filtered off; in plant nutrition, some enzymes can take up specific minerals only in a nondialyzable form.

ꙮ Open-pollination

The pollination of plants by natural methods, such as the transference of pollen from one plant to another by insects or the wind.

ꙮ Parthenocarpic

Describing the production of fruit without the fertilization of plant ovules (i.e. asexually); the resultant fruit are inevitably themselves infertile and, thus, generally seedless, although they may still be propagated by vegetative means.

ꙮ Pepo

Fruit in the form of a berry (a fleshy fruit derived from a single ovary, containing pulp and one or more seeds) but with a hard outer rind; classic examples are melons (and their relatives, cucumbers, and squashes) and papaya, but some would also include passion fruit.

✿ Polyphenols

Chemical compounds comprising phenol units; in fruits such as grapes, which have high polyphenol contents (including tannins (*q.v.*), lignins, and flavonoids (*q.v.*)), it has been suggested (but not proved) that it is the polyphenols that have particularly beneficial effects on human health and well-being.

✿ Pome

Type of fruit of which an apple is characteristic: a thinnish but fairly tough skin encloses fibrous flesh that corresponds to a development of the hypanthium, that in turn surrounds the true fruit which is the core; other pomes include pears, quinces, and loquats, but also hawthorn and rowan berries.

✿ Pseudocarp

Fruit in which the flesh that is eaten is derived not from the plant's ovary but from some other nearby tissue; the fruit of both strawberries and cashew nuts, for example, develop from the "receptacle" that contains the plant's ovaries; also called false fruit (*q.v.*).

✿ Species

Term used to classify, in the greatest detail, an individual type of plant or closely related varieties of plants, that belong to a single genus (*q.v.*). In the text the abbreviations "spp." and "subsp." refer to species (plural) and subspecies respectively, while "var." refers to a division of a species, below the rank of subspecies.

✿ Tannins

Polyphenols (*q.v.*) that appear in plants to control elements of the nitrogen cycle and to be concerned with growth, ripening, and decay; in fruit they may because of their astringent taste also represent a defense against predation by birds and animals.

✿ Zest

The outer, pigmented, oily peel (exocarp (*q.v.*), flavedo) of a citrus hesperidium (*q.v.*), as scraped off (with a zester) and used for culinary purposes, notably to flavor pastries, desserts, cakes, and confectionery or, when cut in spirals, as a cocktail garnish.

Index

Page numbers in **bold type** refer
to color illustrations.

Aegle: bael................68
Alexander the Great.......36
Allwood, John............34
Almond..................10
Amar al din..............18
Anacardium: cashew.....64, 68
Ananas: pineapple...8, 9, 23, 76
Annona: sugar apple.... 84, 85
Appeldoorn, Holland......60
Apple....... 26, 30, 60, 64, 73,
77, 78, 79, 81, 95
Apricot....18, 38, 60, 80, 98, 99
Arbutus: strawberry tree.....11
Arctostaphylos
Bearberry..............30
Manzanita.............30
Artocarpus
Breadfruit........19, 66, 67
Jackfruit................50
Asimina: pawpaw..........77
Avocado..............57, 80

Bacchanalia..............61
Bael....................68
Banana...10, 42, 48, 49, 86, 106
Banana republics..........72
Barbados gooseberry.......82
Bearberry................30
Bible, fruits of the........ 34
Big Apple................43
Bilberry.................86
Black currant............58
Blackberry.... 44, 45, 47, 110
BlackBerry®, the...........93
Blackthorn...............46
Blueberry.......... 87, 88, 89
Boysenberry.............38
Breadfruit.........19, 66, 67
Brix numbers............68
Bush lawyer..............92

Cantaloupe......... 28, 29, 34
Cape gooseberry..........82

Caravaggio, Michelangelo da:
Basket of Fruit............56
Carica: papaya......69, 74, 75
Cashew................64, 68
Celtis: hackberry..........64
Charlevoix, Michigan......69
Cherry......12, 13, 15, 69, 102
Chinchilla, Queensland.....34
Citrullus: watermelon 20, 21, 26,
34, 72
Citrus fruits..............22
Clementine.............43
Kumquat..............111
Lemon........94, 95, 96, 97
Lime.................73
Mandarin.............43
Orange.... 32, 33, 51, 68, 114
Satsuma..............43
Tangerine.............43
Classifying fruit...........73
Clementine...............43
Cleopatra................61
Coconut...........42, 50, 76
Cocos see coconut
Columbus, Christopher 8, 28, 94
Cranberry................ 77
Cucumis
Cantaloupe....... 28, 29, 34
Honeydew melon.....31, 34
Currants..............111
Cyanide.................10
Cydonia: quince...102, 104, 105

Darrow, George M.........38
Dimocarpus: longan........68
Dragonfruit.............23
Drupes.................93
Durian..................14
Durio see durian

Elderberry...............18
Ericaceae family..........86
Eriobotrya: loquat..........46

Ficus see fig
Fig............. 10, 40, 41, 61
Fitz Gerald, John J.........43
Fools (desserts)........... 27
Fortunella: kumquat.......111
Fragaria: strawberry..... 14, 60,
80, 100, 101, 105
Fraser, Mary..............27
Fruit cocktail.............56
Fruit machines............72
Fruitcake................114

Gaultheria: wintergreen....112
Gay, John: [O] Ruddier
than the Cherry.........102
Gaylussacia...............88
Huckleberry........88, 102
Gin....................46
Goji berry...............51
Golden raisins...........111
Gooseberry.......11, 27, 82, 83
Grape..... 10, 19, 34, 110, 111
Grenadine syrup..........106

Hackberry................64
Haskap see honeyberry
Hippophae: sea buckthorn....43
Honeyberry..............114
Honeydew melon.......31, 34
Huckleberry..........88, 102
Hylocereus: pitaya.......... 23

Jackfruit.................50
Juniper..................46
Juniperus see juniper

Kiwifruit............26, 27, 94
Knott, Walter.............38
Kumquat................111

Lemon..........*94, 95, 96, 97*
Lime....................*73*
Limeberry...............*68*
Limequat *111*
Litchi..............*11, 60, 68*
Loganberry.............*23*
Longan.................*68*
Lonicera: honeyberry......*114*
Loquat *46*
Lycium: goji berry *51*

Malus: apple..... *26, 30, 60 64,
73, 77, 78, 79, 81, 93*
Mamoncillo *68*
Mandarin *43*
Mandarinquat *111*
Mangifera see mango
Mango............*68, 90, 91*
   'Tommy Atkins' *76*
Manzanita *30*
Maple *68*
Maple syrup *42*
Martin, Edward...........*43*
Melicoccus: mamoncillo *68*
Melon
   Cantaloupe *28, 29, 34*
   Honeydew............*31, 34*
   Watermelon .*20, 21, 26, 34, 72*
Monster fruit *73*
Monstera see monster fruit
Morus see mulberry
Mulberry *16, 17, 56*
Musa: banana...... *10, 42, 48,
49, 90, 106*
My Love's an Arbutus........*11*

Nectarine *70, 71*
Nephelium: rambutan *68*
Noah....................*34*

Ohio Pawpaw Festival.......*77*
Olea see olive
Olive *14*
Oliver, British Columbia.....*69*
Opuntia: prickly pear.......*27*
Orange *32, 33, 51, 68, 114*
Orange Blossom Special *61*
Orangequat................*111*
Otaheite gooseberry *82*

Pálinka *24*
Papaw *74*
Papaya..............*69, 74, 75*
Passiflora see passion fruit
Passion fruit........... *54, 55*
Pawpaw..................*77*
Peach........*36, 37, 38, 39, 51*
Pear.........*50, 95, 116, 117*
Pereskia: Barbados
   gooseberry..............*82*
Persephone *52*
Phyllanthus: Otaheite
   gooseberry..............*82*
Physalis: Cape gooseberry *82*
Pineapple *8, 9, 23, 76*
Pitaya *23*
Plum *50, 62, 63, 65, 96*
Pomegranate..*35, 42, 52, 53, 106*
Prickly pear *27*
Prunes................... *62*
Prunus................... *96*
   Apricot *18, 38, 60, 80,
                   98, 99*
   Blackthorn *46*
   Cherry.... *12, 13, 15, 69, 102*
   Nectarine *70, 71*
   Peach.......*36, 37, 38, 39, 51*
   Plum *50, 62, 63, 65*
Punica: pomegranate *35, 42,
52, 53, 106*
Pyrus: pear *50, 95, 116, 117*

Quince *102, 104, 105*

Raisins *111*
Rambutan.................*68*
Rameses II *30*
Raspberry... *108, 112, 113, 115*
Red currant *58, 59*
Rhubarb *18, 22, 107*
Ribes
   Black currant *58*
   Gooseberry.....*11, 27, 82, 83*
   Red currant.......... *58, 59*
   White currant...........*58*
Rosa: rose hip *24, 25*
Rosaceae family *35*
Rose hip *24, 25*
Rubus
   Blackberry.... *44, 45, 47, 112*
   Boysenberry *38*
   Bush Lawyer *92*
   Raspberry ..*108, 112, 113, 117*
   Wineberry *112*

Sambucus: elderberry........*18*
Sapindales order *68*
Satsuma *43*
Scurvy...................*23*
Sea buckthorn *43*
Sloe gin *46*
Stenocereus: pitaya *23*
Strawberry *14, 60, 80, 100,
101, 103*
Strawberry tree *11*
Sugar apple........... *84, 85*
Sunquat *111*

Tangerine *43*
Traverse City, Michigan *69*
Triphasia: limeberry.........*68*
Tutti frutti *34*

Vaccinium *88*
   Bilberry.................*88*
   Blueberry......... *87, 88, 89*
   Huckleberry *88, 102*
Vitis: grape ..*10, 19, 34, 108, 109*

Watermelon ...*20, 21, 26, 34, 72*
White currant *58*
Wineberry *112*
Wintergreen..............*110*
Wolfberry................*51*

Picture Credits

The publisher would like to thank the following individuals and organizations for their kind permission to reproduce the images in this book. Every effort has been made to acknowledge the pictures, however we apologize if there are any unintentional omissions.